Contesting Extinctions

Environment and Society

Series Editor: Douglas Vakoch

As scholars examine the environmental challenges facing humanity, they increasingly recognize that solutions require a focus on the human causes and consequences of these threats, and not merely a focus on the scientific and technical issues. To meet this need, the Environment and Society series explores a broad range of topics in environmental studies from the perspectives of the social sciences and humanities. Books in this series help the reader understand contemporary environmental concerns, while offering concrete steps to address these problems.

Books in this series include both monographs and edited volumes that are grounded in the realities of ecological issues identified by the natural sciences. Our authors and contributors come from disciplines including but not limited to anthropology, architecture, area studies, communication studies, economics, ethics, gender studies, geography, history, law, pedagogy, philosophy, political science, psychology, religious studies, sociology, and theology. To foster a constructive dialogue between these researchers and environmental scientists, the Environment and Society series publishes work that is relevant to those engaged in environmental studies, while also being of interest to scholars from the author's primary discipline.

Recent Titles in the series

Contesting Extinctions: Decolonial and Regenerative Futures, edited by Suzanne M. McCullagh, Luis I. Prádanos, Ilaria Tabusso Marcyan, and Catherine Wagner.

Global Capitalism and Climate Change: The Need for an Alternative World System, Second Edition, by Hans A. Baer

Ecological Solidarity and the Kurdish Freedom Movement: Thought, Practice, Challenges, and Opportunities, edited by Stephen E. Hunt

Ecomobilities: Driving the Anthropocene in Popular Cinema, by Michael W. Pesses

Embodied Memories, Embedded Healing: New Ecological Perspectives from East Asia, edited by Xinmin Liu and Peter I-min Huang

Wetlands and Western Cultures: Denigration to Conservation, by Rod Giblett

Sustainable Engineering for Life Tomorrow, edited by Jacqueline A. Stagner and David S. K. Ting

Nuclear Weapons and the Environment: An Ecological Case for Nonproliferation, by John Perry

Portland's Good Life: Sustainability and Hope in an American City, by R. Bruce Stephenson

Conservation, Sustainability, and Environmental Justice in India, edited by Alok Gupta

Environmental and Animal Abuse Denial: Averting Our Gaze, edited by Tomaž Grušovnik, Reingard Spannring, and Karen Lykke Syse

Living Deep Ecology: A Bioregional Journey by Bill Devall, edited with an introduction by Sing C. Chew

Environmental and Animal Abuse Denial: Averting Our Gaze edited by Tomaž Grušovnik, Reingard Spannring, and Karen Lykke Syse

Contesting Extinctions

Decolonial and Regenerative Futures

Edited by
Suzanne M. McCullagh, Luis I. Prádanos,
Ilaria Tabusso Marcyan, and
Catherine Wagner

LEXINGTON BOOKS
Lanham • Boulder • New York • London

Published by Lexington Books
An imprint of The Rowman & Littlefield Publishing Group, Inc.
4501 Forbes Boulevard, Suite 200, Lanham, Maryland 20706
www.rowman.com

6 Tinworth Street, London SE11 5AL, United Kingdom

British Library Cataloguing in Publication Information Available

Library of Congress Cataloging-in-Publication Data Available

ISBN 978-1-7936-5281-2 (cloth)
ISBN 978-1-7936-5282-9 (electronic)

Contents

Acknowledgments

Many were involved in the making of this book. We thank our colleagues at the Myaamia Center, Kara Strass and George Ironstrack, for their transformative participation in discussions of the concept for the volume, and for their inspiring contributions to the symposium that generated it. "Approaching Extinction|Contesting Extinction" was held in March 2020 on Miami University's Oxford, Ohio campus. George and Kara welcomed participants to the traditional lands of the myaamia on behalf of the Miami tribe and gave the opening talk, and Kara guided participants through the remarkable myaamia ribbonwork exhibition she had curated at Miami's Art Museum. Thanks too to Shafkat Khan and the rest of the Changing Climate, Changing Communities research collective for sharing with participants the ecological-education art exhibit they curated at the Oxford Community Art Center. Gratitude to the staff at Circle Bar—the warm, laid-back atmosphere there kept meetings fun.

All of this volume's authors offered thoughtful formal and informal input on the nascent volume and commented fruitfully on other participants' chapters-in-progress. Wesley Leonard's comments at the final symposium discussion session usefully reoriented our thinking about extinction discourses. Rohan Quinby's advice and sharp editorial eye was essential during the final stages of completing the manuscript.

The Humanities Center at Miami University, through director Timothy Melley, provided generous funding for our research collaborative and for the symposium, which was also supported by the Department of English. Athabasca University provided additional support for both the symposium and the volume. We thank the staff at Miami whose work helped make the symposium and volume possible: Lindsay Masters, Elias Tzoc, Neila Hanges, and Rachel Treadway, as well as the redoubtable Polly Heinkl. Our editor at Lexington Books, Kasey Beduhn, has been a source of cheer and swift assistance whenever we have needed it. Special thanks to Christopher Vanbeukering for helping us with referencing at the eleventh hour, and to Evan Supple for creating the index. Not least, we thank our human and non-human animal family members for their patience and love: Ambrose, Strawberrius, Sonny, Shiva, Shakti, Devi, Daniela, Luna, and Mitra WillowWolf Blue.

Introduction

Suzanne M. McCullagh, Luis I. Prádanos, Ilaria Tabusso Marcyan, and Catherine Wagner

We write amidst accumulating narratives of extinction. In scientific, literary, and speculative writings, we find warnings of an imminent mass extinction event brought on by anthropogenic climate disturbances and drastic changes in land use related to capital's urbanization of the planet. Global warming, resulting from rising greenhouse gas emissions brought about by burning fossil fuels, deforestation, and industrial agriculture, is undermining the conditions for many forms of life on the planet. *Contesting Extinctions* attends to this contemporary moment, examining narratives of extinction not as neutral accounts but as expressions of agendas, ideologies, and imaginaries that exert uneven effects on different human and nonhuman communities and their habitats. The extinction narratives that become dominant—as well as those that are erased—will shape political imaginations and condition the solutions proposed and the responses considered. How extinctions are conceived influences political, economic, cultural, and technical responses that will have profound consequences for the planet and the living beings who inhabit it.

Contesting Extinctions critically interrogates the ways that extinctions are discursively framed, as well as the systems that bring about biocultural loss. As a decolonial project, it "questions the 'civilization of death' hidden under the rhetoric of modernization and prosperity" and affirms "the principle that the regeneration of life shall prevail over primacy of the production and reproduction of goods at the cost of life."[1] To contest extinction is to contest and oppose coloniality's "propagation of capitalism, racism, the modern gender system, and the naturalization of the death ethics of war."[2] We distinguish the decolonial practice of delinking from colonial systems of knowledge and social and ecological organization from decolonization, which, following Eve Tuck and K. Wayne Yang, "specifically requires the repatriation of Indigenous land and life."[3]

We begin our introduction with "Contesting Extinction Narratives," where we examine the quantitative accounts of species extinction in scientific discourse and argue that these studies, while claiming to call attention to imminent biological annihilation, fail to question the colonial, capitalist, and patriarchal systems that produce this predicament. The next section, "Contesting the 'Anthropocene,'" turns to the problematic resonances between extinction discourse and the concept of the Anthropocene insofar as they posit universal humanity as the agent of climate change and biodiversity loss, thus obscuring the differential responsibilities of those invested in and profiting from settler-colonial, industrial, and extractivist relations with land. "Contesting Extinction Temporalities" analyzes the temporal logics at work in extinction discourses and the risks of their uncritical deployment. "Contesting Extinctions through Critical Relationality" suggests that scholars (and others) take up a generatively critical attitude toward the socioecological relations in which they participate, especially as the dominant relational paradigm is fatally dangerous to species life and cultural persistence. We end the introduction with "Contesting Extinctions as a Collaborative Practice," which tells the story of this volume's conceptual unfoldings and introduces the chapters.

CONTESTING EXTINCTION NARRATIVES

A plethora of scientific writings describes the massive biodiversity loss that threatens the viability of all biological life on the planet. The 2018 *WWF Living Planet Report* "shows that population sizes of wildlife decreased by 60% globally between 1970 and 2014."[4] The 2019 IPBES global assessment on biodiversity and ecosystem services,[5] the most comprehensive publication to date on the situation of the planetary living systems, depicts the drastic acceleration of biodiversity loss during the last decades. A 2017 article in *PNAS* quantitatively analyzes the population losses and declines in vertebrates and concludes that the current rates of decline entail a process of "biological annihilation."[6] Similar conclusions have been drawn by other scientists in relation to drastic declines in the biodiversity of insects, plants, and soil microorganisms. Regarding aquatic ecosystems, the situation is not encouraging either: approximately 75% of large rivers are not free-flowing and more than half of oceans are degraded.[7] This scenario is likely even worse than it appears when we consider "co-extinction" processes through which the loss of one species triggers the decline of other interrelated species.[8] The rate of biodiversity loss presents an imminent risk for the systemic collapse of the living systems that make human existence viable.

In January of 2021, a group of concerned environmental scientists published a brief report that unambiguously summarized the urgency of the situation and encouraged immediate, effective, and drastic societal changes. The title, "Understanding the Challenges of Avoiding a Ghastly Future," indicated the authors' intention to explain the "overwhelming challenges ahead and 'tell it like it is.'" The authors seemed frustrated by the fact that "[t]he science underlying these issues is strong, but awareness is weak," and lamented that "[t]he scale of the threats to the biosphere and all its life-forms—including humanity—is so great that it is difficult to grasp for even well-informed experts."[9] They worried that dominant economic and political systems are ill-prepared and perhaps completely incapable of averting catastrophe.

All these studies point to a clear conclusion: the current trend cannot be maintained without collapsing ecological systems that plants and nonhuman and human animals depend upon for survival. However, these studies say nothing about the historical contingencies and asymmetrical power relations that brought *us*, "the denizens of the Earth,"[10] to the brink of extinction, nor do they point to the differential responsibilities for the design and global propagation of an economic metabolism that destroys the conditions for life. In other words, these warnings fail to address the colonial,[11] patriarchal,[12] and imperial processes[13] that have brought us where we are. Worse yet, as Noah Theriault and Audra Mitchel argue, mass extinction narratives, by

> eliding the violent structures that disproportionately burden certain assemblages of beings . . . naturalize a colonial order in which some earthlings are actively targeted for extermination, some are categorized as valuable "biodiversity," and many others are summarily consigned to an unmarked planetary grave.[14]

As such, most proposed solutions do not target the real roots of our social and ecological problems.

Since its inception, capitalism has worked to extinguish relational ways of living and understandings of land as commons.[15] From the enclosure acts in early modern Europe that privatized lands held in common, the abduction and enslavement of African peoples, and the continued dispossession of Indigenous peoples around the world, the capitalist drive for growth has from its beginning exterminated different life-worlds. As Vandana Shiva points out,

> [t]he transformation of the commons into commodities . . . deprives the politically weaker groups of their right to survival, which they had through the commons, and it robs from nature its right to self-renewal and sustainability, by eliminating the social constraints on resource use that are the basis of common property management.[16]

The expansion and intensification of an economic culture addicted to growth not only transforms life into infrastructure and commodities[17] but accumulates massive economic and political power in the hands of a few while exploiting and impoverishing the majority of the human population. In the words of ecofeminist philosopher Val Plumwood,

> [t]here is clearly a serious problem about the ecological rationality of a system that allows those who have most access to political voice and decision-making power to be also those most relatively remote from the ecological degradation it fosters, and those who tend to be least remote from ecological degradation and who bear the worst ecological consequences to have the least access to voice and decision power.[18]

Extinction and inequality are co-constitutive Capitalocene[19] processes that need to be considered in tandem. As Arturo Escobar puts it, techno-industrial modernity "can be considered a massive design failure" and, as with any other way of organizing socioecological relations that is not pursuing "zero poverty and zero extinction," should be considered unacceptable.[20] We (the editors of this volume) wish to make unambiguously clear our belief that attempts to solve our social and ecological issues that do not aim at decolonization are going to make things worse. As the Red Nation has declared in *The Red Deal*, "Ending settler colonialism and capitalism and returning Indigenous lands are all possible—and necessary. . . . We need a revolution of values that recenters relationships to one another and the Earth over profits."[21] For the Red Nation, the "path forward is simple: it's decolonization or extinction."[22] In other words, socially desirable and ecologically viable futures are also decolonial futures.

The depletion of the Earth's (re)generative capacities is inseparable from social-ecological inequities and environmental racism. As Indigenous peoples, environmental justice activists, and scholars have long known, racism, patriarchy, and settler colonialism are at the root of the climate crisis, with the result that Indigenous, Black, and other people of color disproportionately suffer the effects of climate change and ecosystem destruction. Within these groups, women and children are more intensively affected by environmental changes and conflicts. The coauthors of "The Snarled Lines of Justice: Women ecowarriors map a new history of the Anthropocene" assert that

> [t]oday there can be no more trust in hiding places . . . no more deflections from the structural inequalities we manufactured to invent "safety" for some and danger for others. Some of us have known this for a while. . . . Now most everyone can taste the fumes of the Anthropocene and the planetary legacies of invasion, extraction, and exploitation that have poisoned and sickened us.[23]

As the ecological consequences of capitalism become increasingly apparent, it is imperative to attend to and reckon with the inequalities that allow some to maintain positions of safety while disavowing responsibility for the dangerous ecological conditions that others are subjected to.

According to Raj Patel and Jason Moore, capitalism thrives by putting nature to work as cheaply as possible: "Cheap is a strategy, a practice, a violence that mobilizes all kinds of work—human and animal, botanical and geological—with as little compensation as possible."[24] These mobilizations cause extinction, which can be understood as an obvious outcome of the capitalist cheapening strategy. As Justin McBrien explains, "[t]he logic of accumulation is not capable of outrunning extinction because accumulation and extinction are the same process. They cannot be decoupled. But the human being can be decoupled from Capital. Capital is extinction. We are not."[25]

Current extinction trends are the result of homogenizing and universalizing (patriarchal, colonial, racist, growth-oriented) power relations in which differences are translated into inequalities. Contesting ongoing culturally and biologically impoverishing inertias would entail embracing and inhabiting the pluriverse,[26] a world where many worlds fit, as the Zapatistas say. When the dominant cultural paradigm translates differences into inequalities that serve extractivist agendas, the result is rolling extinctions of multiple interconnected lifeways.

Under what has been named (as an alternative to "Anthropocene") the Capitalocene,[27] the intensification of life-capital conflict tends to be resolved in favor of capital. When human social reproduction is organized around capital accumulation—which is a powerful engine for extraction—anthropogenic extinction is the result. However, if societies decide to favor life over capital and organize social reproduction according to ecofeminist principles or traditional Indigenous practices of land stewardship—ethics of care—our societies could function as life-enhancing systems. In Plumwood's words,

> [w]e must counter those maladaptive forms of reason that radically distance us from the non-human sphere and disguise or disappear our ecological embeddedness and vulnerability, in order to develop a communicative, place-sensitive culture which can situate humans ecologically and nonhumans ethically.[28]

There is nothing inherently human to blame for triggering and exacerbating extinctions and inequalities. As Ashley Dawson argues, "extinction needs to be framed not as a product of some general human capacity for despoliation of the planet . . . but rather as the product of a global attack on the commons" by capitalist interests.[29] If debates about the Earth's carrying capacity abandoned their Malthusian focus on the number of humans, researchers could pay fruitful attention to the diverse social metabolisms of specific organizations—that is, the energy and material intensity mobilized to satisfy needs in a given social system.[30] When humans labor under paradigms that demand growth, their social metabolism becomes increasingly expansive and destructive, generating inequalities and extinctions. Conversely, humans organized around regenerative cultural practices and ecofeminist ethics may display a beneficial metabolism that enhances human well-being while improving ecosystemic health. Empowered women with access to family planning and control over their reproductive decisions tend to have smaller families, so in such a system, human populations would likely not grow unchecked.[31] There is also a historical correlation between the energy intensification of techno-industrial modernity and population growth.[32] A less energy-intensive social metabolism is unlikely to trigger explosive population growth. In short, societies that encourage individualistic, competitive, and consumerist mindsets tend to deplete the planet and generate corrosive social environments, while populations practicing permaculture principles and frugal abundance have the potential to heal communities, soils, and ecosystems. Humans can enhance the likelihood of long-term social reproduction on a finite planet by promoting post-growth and pluriversal imaginaries as well as embracing regenerative and convivial practices.[33] Now is the time to contest extinctions by thinking and acting to dismantle the unjust power relations, institutional arrangements, and distorted social imaginaries behind the dominant necrotic social and ecological trajectory. It's time to support and nurture economic metabolisms, ways of living, and social relations that facilitate social-ecological regeneration.

CONTESTING THE "ANTHROPOCENE"

The concept of the Anthropocene, which began as a debate among geologists (and is still ongoing) about whether to categorize evidence of human activity in the geological record as a specific geologic epoch, has gone mainstream. Popular attention to anthropogenic causes of the present ecological predicament is welcome and necessary. Unfortunately, by seeming to propose humans in the aggregate as agents of ecological destruction, the concept of the Anthropocene recycles assumptions that facilitate the extractive processes responsible for the crisis. By rendering invisible the ways that *some* humans dominate and oppress other humans and nonhumans, the Anthropocene concept fails to call into question the ontological, ethical, and political problems of human domination of the more-than-human world.

Recently, scholars have been contesting the Anthropocene concept for insufficiently differentiating and recognizing the specific social organizations and modes of living that bring about social and ecological destruction. Zoe Todd notes that not all humans "are equally implicated in the forces that created the disasters driving contemporary human-environmental crises" and argues that "not all humans are equally invited into the conceptual spaces where these disasters are theorized or responses to disaster formulated."[34] Todd reads the concept of the Anthropocene as inattentive to, even silencing of, non-White and non-European voices and ways of living. Todd and Heather Davis argue that settler colonialism's uneven impacts are erased by Anthropocene concepts that do not attend to the relationships of Indigenous peoples with land and other-than-human beings—relationships that settler colonialism, both past and present, undermines. "[S]ettler colonialism, and its extensions into contemporary petrocapitalism," write Todd and Davis, "is a severing of relations. It is a severing of relations between humans and the soil, between plants and animals, between minerals and our bones. This is the logic of the Anthropocene."[35]

Moore contends that the concept of the Anthropocene, by naming "humanity as an undifferentiated whole," fails to locate "the origins of the modern geohistorical shift" in terms of class, capital, imperialism, and/or culture.[36] He prefers the word Capitalocene, thereby naming capitalism as the locus of agential responsibility. Eileen Crist argues that by

> affirming the centrality of man—as both causal force and subject of concern—the Anthropocene shrinks the discursive space for challenging the domination of the biosphere, offering instead a techno-scientific pitch for its rationalization and a pragmatic plea for resigning ourselves to its actuality.[37]

While Crist worries that the Anthropocene concept "crystallizes human dominion,"[38] Sylvia Wynter usefully complicates the account of the "human" at work in such discourses. Naming human activities as the cause of climate change and species loss, according to Wynter, constrains thinking within the epistemic terms of "Western and mimetically Westernized middle classes" and the "master discipline of economics."[39] Wynter argues that *homo oeconomicus*' mode of being human has been falsely elevated to the status of a universal, and equated with all of humanity. Free-market capitalism, as the model of social organization befitting to *homo oeconomicus*, can "be enacted only on the homogenized basis of the systemic repression of all other alternative models of material provisioning."[40] She argues that if the "'we' who are destroying the planet . . . are not understood as the referent-we of *homo oeconomicus*" then the proposals for change that will be proposed "are going to be devastating!"[41]

Similarly, in *Forces of Reproduction*, Stefania Barca combines ecofeminist thought and historical materialism to argue that the dominant narrative of the Anthropocene is only the latest version of the capitalist/industrial/colonial/patriarchal mainstream discourse of Western modernity that "identifies humanity with the male master of 'nature' . . . which assumes the master model of modernity as representing the entire human species, and den[ies] the existence or historical relevance of non-master agencies and possibilities."[42] For Barca, this narrative not only takes Anthropocentrism to a new level of human geo-supremacy by celebrating, once again, the forces of production of industrial modernity, but it also does not account for the forces of reproduction, that is, the "subjects who reproduce humanity by taking care of the biophysical environment that makes life itself possible . . . and keep the world alive."[43] The Anthropocene narrative implies that the solution to the environmental crisis is to amplify and intensify the techno-industrial master model that caused it. A *techne* that has justified and triggered inequality and rolling extinctions during the last few centuries should not be deployed to frame the crisis or to design its solution. That is why Barca claims that "keeping the world alive requires dismantling the master's house."[44] Again, the dilemma is not about overcoming a bad Anthropocene with a good Anthropocene (good for what and for whom?) but about a choice between decolonization and multiple, interlocking extinctions. Decolonial epistemologies thus entail contesting Anthropocene neocolonial and patriarchal frames and stories.

CONTESTING EXTINCTION TEMPORALITIES

Although the idea of the Anthropocene, among other extinction narratives, may rouse people to act on the climate crisis, deploying it runs the risk of reproducing and amplifying settler colonialism and racial capitalism. Kathryn Yusoff draws attention to the configuration of the Anthropocene in the "future tense rather than in recognition of the extinctions already undergone by black and indigenous peoples"[45] and argues that the practices of "mining, extraction, waste, extinction" that constitute the Anthropocene "are the product of, and reinforce, colonial divisions of power, territory, and life."[46] The concept of extinction has been and continues to be used materially and discursively to erase the continued existence of Indigenous peoples, as well as the viability of Indigenous modes of living and relating to the land (including Indigenous languages). As extinction discourses proliferate as ways to frame and grapple with our ecological moment, we might pause to examine the temporal logics undergirding these discourses and attend to the political and ecological consequences of their uncritical deployment.

Extinction discourses tend to participate in a temporal logic that deprives those in the presence of agency. The concept of extinction operates in the present by way of a double erasure. First, it relegates peoples, languages, and species into the past. The declaration that an Indigenous people or language is extinct, while frequently expressed in the form of a lament for a lost past, discursively erases Indigenous peoples in the present. While we do not deny that species and lifeways can be and are being irretrievably lost, it is important to note that labeling a group or language "extinct" is not a neutral act. By relegating capitalist social-ecological relations and settler-colonial agency to the past, it obscures their effects in the present, thereby perpetuating them into the future, and undermining the political power and cultural presence of Indigenous peoples. The erasure of Indigenous history and presence cuts the social imaginary off from a diversity of possible and abundant futures.

Second, the concept of extinction is often used to name the "end of the world" as a future event toward which "we" are all heading. When deployed in this manner, the concept of extinction erases from the social imaginary the awareness and recognition that many peoples (especially Indigenous and Black) have already experienced and survived multiple extinctions of their worlds.[47] While reflecting on the extinctions of plants, nonhuman animals, ecosystems, and the lifeways entangled with them is a novel activity for many whose concerns have been activated by the climate crisis, there are others for whom grappling with the end of the world is not new. Kyle Powys Whyte and Kim TallBear assert that anxiety about the future among settler subjects is a familiar condition for Indigenous subjects. In their consideration of contemporary apocalyptic concerns, both thinkers expose the temporal multiplicity and nonlinearity (rather than Euro-modern temporal homogeneity and linearity) that is covered over by frequent invocations of extinction and the Anthropocene as future events. Whyte explains that "some indigenous peoples already inhabit what our ancestors would have likely characterized as a dystopian future." Settlement in the nineteenth and twentieth centuries forcibly severed many Indigenous peoples' "relationships with plants, animals and ecosystems at a wrongfully rapid pace." Anthropogenic climate change and its impacts—such as "changes associated with deforestation, forced removal and relocation, containment on reservations (i.e., loss of mobility), liquidation of [Indigenous] lands into individual private property and subsequent dispossession, and unmitigated pollution and destruction of [Indigenous] lands from extractive industries and commodity agriculture"—have already been suffered by many Indigenous peoples "at the hands of settlers." Whyte points out that these anthropogenic environmental changes "refer specifically to how industrial settler campaigns both dramatically changed ecosystems . . . and obstructed indigenous peoples' capacities to adapt to the changes."[48] TallBear clearly asserts:

> This is not our apocalypse. . . . My apocalyptic memory goes back to [the] 1862 . . . partitioning of Dakota lands in order to eradicate Dakota social

> formations. . . . The goal was elimination . . . and we are still rebuilding from the quieting of Indigenous worlds.

Those who have "not found shelter within that progressive colonial narrative" are not experiencing the "same kind of diminished ontological hope"[49] that contemporary apocalypse narratives evoke for some. In this sense, contesting extinctions means (re)opening and reclaiming the possibility for collective and relational futures beyond—and radically different from—the unequal, extractive, and exploitative relations ingrained in colonialism.

CONTESTING EXTINCTIONS THROUGH CRITICAL RELATIONALITY

Because relations are contingent on particular historical and material situations, they have the potential to be reoriented and changed. The social and economic forces that currently structure relations among many of the planet's inhabitants were formed under settler-colonial and imperial regimes. These regimes encouraged and enforced extractive relationships enacted through manufactured hierarchies that classified, naturalized, and exploited differences. When we open our thinking toward counterhegemonic relational paradigms (for instance, if we view more-than-humans as potential kin[50] rather than as objects for exploitation), alternative structures such as non-extractive systems of social reproduction become available. Critical relationality, a term we borrow from Kim TallBear and Angela Willey, describes an approach that "center[s] in our thinking the material conditions of possibility for our own constrained choices and the distribution of harms and benefits in which they are imbricated."[51] For example, if we view productivity per acre in terms that value workers, soils, and communities, we see that it is higher in nonindustrial farming. More people are fed clean nutritious food, and soil microorganisms thrive. When we view the same acreage in terms of the desire of a landowner to reap monetary profits from exploited land and labor, industrial farming appears to be cheaper, because this accumulation process externalizes the cost of labor and doesn't account for ecological damage.[52] In addition, industrial farming devalues and erases traditional (Indigenous) knowledges that enable soil regeneration and can produce higher yields.[53] A critical orientation toward ongoing settler-colonial practices of land tenure might lead more food providers to investigate the reciprocal relationships with land and soil practiced by Indigenous peoples on Turtle Island and elsewhere.

We believe that critical relationality is a useful approach for "contesting extinctions." A core claim here is that "extinction" happens in relation to agencies and powers. We don't just live in relation to the cessation of species life or linguistic persistence—we participate in relations that *lead to* these cessations. As TallBear and Willey write, "[o]ur ability to imagine nature and relationality differently are deeply enmeshed, and this imaginative work is critical to the reworlding before us."[54] We propose acting in critical relation—not just to the practices and powerful agents that bring about rolling extinctions—but to "extinction" as a concept whenever it is mobilized and articulated without confronting the dominant imaginary that justifies extractive practices and power asymmetries. Critical relationality helps us grapple with the material and rhetorical implications of "extinctions," and encourages us to examine our own relationships with human and more-than-human actors as a way of learning to be in better relation—not asserting teleologies, but making kin with our existing relatives toward a regenerative future.

CONTESTING EXTINCTIONS AS A COLLABORATIVE PRACTICE

The practice of critical relationality implies taking a close look at our own practices of relating. In planning both symposium and this volume, our editorial collective sought to engage in a porous, collaboratively driven process that could swerve in response to input. This volume, therefore, has taken shape in a spirit of collaboration.

We formed the Environmental Humanities Research Collaborative at Miami University in the summer of 2018 as an extension of the ideas fostered during "Urban Futures," a Miami University Humanities Center Altman Program hosted in 2017–2018. On Friday afternoons, we met at an obscure bar frequented mainly by locals and situated in a back alley in Oxford, Ohio. Together in the Circle (the bar's name), we drank Truth (a local IPA) and discussed readings selected by each of us in turn. Our meetings generated not only the idea for a symposium on "Approaching Extinction" but mutual trust, much laughter, and a collaborative and dialogic ethic we have tried to bring forward into the present volume.

With help from our collaborators—especially from Myaamia tribe members George Ironstrack, Kara Strass, and Wesley Y. Leonard, who generously critiqued our assumptions and terms—we reenvisioned our project multiple times, first changing the name of the symposium from "Approaching Extinction" to "Approaching Extinction | Contesting Extinction," finally moving to the plural with our volume title "Contesting Extinctions." The plural is intended to convey two things: first, that "extinction" is not a singular shared event, but a process that continues to affect different populations and living beings unevenly across time and space. Second, as we have discussed, rhetorics of extinction are not neutral in their orientation toward populations facing genocide or destruction. Extinctions are agential practices: "extinguere," the Latin verb from which the noun "extinction" derives, implies an active subject and an object acted upon. By pluralizing "extinction" in our volume title, we hope to draw critical attention to the term as a rhetorical device used to reaffirm dominant extractive practices by naturalizing and de-agentializing processes of extinguishment.

As editors, we sought to cultivate a collaborative, transformatively dialogic process, not just between editors and authors, but between chapter authors. Chapters were read by editors and authors prior to their presentation at the *Approaching Extinction | Contesting Extinction* symposium held at Miami University (Ohio) on March 2–3, 2020. The symposium concluded with a workshop at which editors and authors provided feedback on the chapters, discussed emerging themes, and co-created the focus for the volume. There was significant discussion about the problematic term "extinction." We all agreed, however, that avoiding the use of the term would be to allow dominant extinction narratives to go uncontested. A major outcome of this conversation was the formulation of the volume title, *Contesting Extinctions: Decolonial and Regenerative Futures*. The collaborative process has resulted in a transdisciplinary collection in which each contribution is in conversation with the others, implicitly acknowledging that, as mushrooms and other biotic systems teach us, our lives and work are co-constituted and mutually transformed.

The symposium was, for us, the last major in-person event before COVID-19 forbid physical contact and marked the beginning of "remote" relating and collaborating. The global pandemic has shed new light on social, racial, and economic inequities that disproportionally affect members of disadvantaged communities, particularly Black, Latinx, and Indigenous people.[55] It has also galvanized a new era of activism among those communities and others and launched a new era of multiracial protest and alliance toward change. In that regard, the pandemic was a generative context for writing and editing a collection that explores racialized and settler-colonial extinction discourses and articulates alternative, regenerative realities, and potentialities. In its final form, *Contesting Extinctions* acknowledges the diverse histories and temporalities of the extinctions faced by nonhuman and human subjects, challenges the idea that the future can be managed by specialists, and disputes dominant narratives about the inevitability of extinction.

Narratives of extinction depend on claims—implicit or explicit—about what has caused the destruction of lives and lifeways. Our book's contestations of such narratives must be grounded in a convincing account of the practices and ethics that promote planetary sustainability. The first chapter provides that account. In "Decolonize, ReIndigenize Planetary Crisis: Biocultural Diversity, Indigenous Resurgence and Land Rematriation," Leonardo E. Figueroa Helland, Abigail Perez Aguilera, and Felix Mantz show that planetary sustainability depends on ensuring Indigenous communities' ongoing reciprocal relationships with the land. Despite ongoing land theft and exploitation, Indigenous land practices foster ecosystem richness and regeneration wherever they continue to exist. Figueroa, Perez, and Mantz explain that defending and revitalizing Indigeneity while dismantling colonial and associated power structures will be essential to overcoming crises of extinction and "p/refiguring" alternatives to the systems that drive them.

In chapter 2, Ryan Heryford explores a poetic dismantling of a colonial extinction narrative: M. NourbeSe Philip's book-length poem *Zong!* The poem is titled after the 1781 slave ship *Zong*, whose captain ordered over 150 African people to be cast into the sea to collect insurance on the "property." Heryford's chapter highlights *Zong!*'s radical attempt to recover and recuperate voices that cannot be translated into available semantic paradigms of preservation and loss—a conundrum that Philip describes as "the inability to tell the story that must be told"[56] and that Édouard Glissant called "a clamor for the right to opacity."[57] Heryford's reading engages a range of decolonial and diasporic thinkers, including Black studies scholars Aimé Césaire, Saidiya Hartman, and Fred Moten, who turn to the Zong massacre as a pivotal moment in ontological constructions of what Sylvia Wynter calls the human as "Man."

Chapter 3, "Philosophizing Extinction: On the Loss of World and the Possibility of Rebirth through the Language of the Sea," similarly reads the sea as a zone of discursive possibility. Marjolein Oele evaluates philosophical conceptualizations of our anthropogenically damaged oceans—zones of loss that connect disparately experienced and ecologically threatened homeplaces across geographies and species. The work of Peter Sloterdijk, Michel Serres, Karin Ingersoll, and Édouardo Glissant, as well as Oele's own experience as an island-dweller from Zeeland, inform her argument that experiences of shared loss can foster a multilinguistic and nonhierarchical ethics and politics, and even a language responsive to the sea.

Lisa Ottum's "What We Talk About When We Talk About Extinction" (chapter 4), like Oele's contribution, also interrogates affective dimensions of extinction narratives. Reading novelist Lydia Millett alongside Deborah Bird Rose, Glenn Albrecht, Lee Edelman, and others, Ottum makes a case for a queer ecological love that foregoes the mourning of human reproductive futurity in favor of a new stance toward extinction. Love in the contested "Anthropocene" can be founded, for Ottum, not so much on dyadic attachments between discrete entities as on interlinked, interspecies systems of symbiosis that tend away from life-destroying systems such as capitalism and colonialism. Such attachments could help us "envision the possible futures surrounding extinction in terms of abundance and desire, rather than (only) scarcity, fear, and anxiety."

Alex Benson's contribution (chapter 5) works at the intersections, both discursive and material, between zoological study and settler colonialism in North America. "Rat-fall: Time and Taxa in the Colorado River Delta, c. 1900" opens out from a discussion of a 1901 short story to draw connections between the nineteenth-century zoology and racialized labor and the twentieth-century settler infrastructures that have damaged the ecology of Western Turtle Island waterways. The intriguing ways in which Oskison, a citizen of the Cherokee Nation, has transformed his scientific source material leads Benson to interpret the story not only as an unsettling of the commonplace analogy between species extinction and the genocidal trope of the "vanishing Indian," but as a queering of colonialist conceptualizations, among them settler time (linear, teleological) and settler taxonomies (which serve to prescribe and limit both the biotic and the political).

The final chapter, by linguist Wesley Y. Leonard, brings us full circle as an instance of the kind of research Figueroa, Perez, and Mantz call for in the first chapter—community-embedded scholarship that resists colonizing definitions and narratives and defends and revitalizes Indigenous lifeways. "Contesting *Extinction* through a Praxis of Language Reclamation" is an account of the reclamation of myaamia, the language of the Miami people, as a form of decolonial restoration work. Language reclamation praxis resists colonial framing of myaamia as "extinct" (a colonial logic of erasure that aligns with discourses of Indigenous "disappearance"). Leonard, an enrolled Miami tribe member, explains that language reclamation praxis draws from and facilitates a critical relationality; in a frame of reclamation, language work can occur as part of a broader response to cultural, political, and spiritual ruptures while supporting community wellness, healthy relationships, and regenerative futures.

CONTESTING EXTINCTIONS AND FOSTERING REGENERATIVE FUTURES

Contesting extinctions means recognizing that linear and universalist narratives—from the celebrated end of history to the "Anthropocene" story—favor hubristic (neo)colonial and neoliberal thinking that permits little room for collective, culturally differentiated political imaginations that can respond with dignity to inequitable and necrotic socioecological trajectories. Failing to contest extinctions would mean giving up the possibility for viable, desirable futures, as well as succumbing to the neoliberal assumption that there is no alternative to the manufactured inequalities and social, cultural, and biological death-regimes that colonial, patriarchal, growth-oriented ideologies have set in motion. Is there nothing left to do but witness the daily disappearance of species and the oppression and extermination of diverse epistemologies and livelihoods? This volume recognizes and explores other possible paths by invoking stories, critiques, and prophecies that are not only emancipatory and empowering but point to socially desirable and ecologically viable futures for all.

NOTES

1. Walter Mignolo, "Epistemic Disobedience, Independent Thought and Decolonial Freedom," *Theory Culture & Society* 26, no. 7–8 (2009): 3.

2. Nelson Maldonado-Torres, "Thinking through the Decolonial Turn: Post-continental Interventions in Theory, Philosophy, and Critique—An Introduction," *TRANSMODERNITY: Journal of Peripheral Cultural Production of the Luso-Hispanic World* 1, no. 2 (Fall 2011): 1.

3. Eve Tuck and K. Wayne Yang, "Decolonization is Not a Metaphor," *Decolonization: Indigeneity, Education & Society* 1, no. 1 (2012): 21.

4. M. Grooten and R. E. A. Almond, *Living Planet Report—2018: Aiming Higher*, WWF, Gland, Switzerland, (2018): 90, https://www.wwf.org.uk/sites/default/files/2018-10/wwfintl_livingplanet_full.pdf.

5. E.S. Brondizio, J. Settele, S. Diaz and H. T. Ngo, "Global Assessment Report on Biodiversity and Ecosystem Services of the Intergovernmental Science-Policy Platform on Biodiversity and Ecosystem Services," *IPBES* (2019): 1, https://doi.org/10.5281/zenodo.3553579.

6. Gerardo Ceballos, Paul R. Ehrlich, and Rodolfo Dirzo, "Biological Annihilation Via the Ongoing Sixth Mass Extinction Signaled by Vertebrate Population Losses and Declines," *PNAS* 114, no. 30 (2017), https://www.pnas.org/content/114/30/E6089.

7. G. Grill, B. Lehner, and M. Thieme, et al., "Mapping the World's Free-Flowing Rivers," *Nature* 569 (2019), https://doi.org/10.1038/s41586-019-1111-9.

8. Giovanni Strona and Corey J.A. Bradshaw, "Co-extinctions Annihilate Planetary Life During Extreme Environmental Change," *Scientific Reports* 8, no. 16724 (2018), https://doi.org/10.1038/s41598-018-35068-1.

9. Corey A. J. Bradshaw et al., "Understanding the Challenges of Avoiding a Ghastly Future," *Frontiers in Conservation Science* (13 January 2021): 1–10, https://doi.org/10.3389/fcosc.2020.615419.

10. Mick Smith, *Against Ecological Sovereignty: Ethics, Biopolitics, and Saving the Natural World* (Minnesota: University of Minnesota Press, 2011), 136.

11. Kathryn Yusoff, "The Inhumanities," *Annals of the American Association of Geographers* 111, no. 3 (2021): 663.

12. Imani Perry, *Vexy Thing: On Gender and Liberation* (Durham: Duke University Press, 2018).

13. Rob Nixon, *Slow Violence and the Environmentalism of the Poor* (Cambridge: Harvard University Press, 2011) and Simone Bignall and Daryle Rigney, "Indigeneity, Posthumanism and Nomad Thought: Transforming Colonial Ecologies," in *Posthuman Ecologies: Complexity and Process after Deleuze*, ed. Rosi Braidotti and Simone Bignall (New York: Rowman & Littlefield, 2019), 159–181.

14. Noah Theriault and Audra Mitchell, "Extinction," in *Anthropocene Unseen: A Lexicon*, ed. Cymene Howe and Anand Pandian (Punctum Books, 2020), 177.

15. According to David McNally, "[f]or capitalism to develop, customary ties between people and the land must be severed, and communal obligations among people disrupted." David McNally, *Monsters of the Market: Zombies, Vampires and Global Capitalism* (Boston: Brill, 2011), 37. For further commentary on enclosure acts and the breaking down of communal obligations, see: Silvia Federici, *Witches, Witch-Hunting, and Women* (Oakland: PM Press, 2018), 21.

16. Vandana Shiva, *Earth Democracy* (Cambridge: South End Press, 2005), 29.

17. Emily Elhacham, et al., "Global Human-Made Mass Exceeds All Living Biomass," *Nature* 588 (2020): 442–444, https://doi.org/10.1038/s41586-020-3010-5. In other words, the globalization of a growth-oriented economic metabolism is transforming planetary life into "modern" infrastructure.

18. Val Plumwood, "The Crisis of Reason, the Rationalist Market, and Global Ecology," *Millennium: Journal of International Studies* 27, no. 4 (1998): 923.

19. Jason W. Moore, ed., *Anthropocene or Capitalocene? Nature, History, and the Crisis of Capitalism* (Oakland: PM Press, 2016), 116–137.

20. Arturo Escobar, *Designs for the Pluriverse: Radical Interdependence, Autonomy, and the Making of Worlds* (Durham: Duke University Press, 2018), 33 and 200.

21. The Red Nation, *The Red Deal: Indigenous Action to Save Our Earth* (Brooklyn: Common Notions, 2021), 13.

22. The Red Nation, *The Red Deal*, 7.

23. Janelle Baker, et al., "The Snarled Lines of Justice. When Women Warriors Map a New History of the Anthropocene," *Orion Magazine* (November 20, 2020), https://orionmagazine.org/article/the-snarled-lines-of-justice/?mc_cid=bf6614eb60&mc_eid=cdbabfa945.

24. Raj Patel and Jason W. Moore, *A History of the World in Seven Cheap Things: A Guide to Capitalism, Nature, and the Future of the Planet* (Berkeley: University of California Press, 2017), 19, 22.

25. Justin McBrien, "Accumulating Extinction: Planetary Catastrophism in the Necrocene", in *Anthropocene or Capitalocene?* ed. Jason W. Moore (Oakland: PM Press, 2016), 135.

26. Mario Blaser and Marisol de la Cadena, "Introduction: Pluriverse Proposals for a World of Many Worlds," in *A World of Many Worlds* (Durham: Duke University Press, 2018), 4.

27. Moore, *Anthropocene or Capitalocene?* 116–137.

28. Plumwood, *Environmental Culture: The Ecological Crisis of Reason* (New York: Routledge, 2002), 239.

29. Ashley Dawson, "Biocapitalism and De-extinction," in *After Extinction*, ed. Richard Grusin (Minneapolis: University of Minnesota Press, 2018), 175–176.

30. "[T]he framework of population has activated and continues to activate racisms, colonialisms, nationalisms, and heteropatriarchies so intensely that it is not a useful framework toward building reparative or decolonial politics toward better worlds." Marilyn Strathern, Jade S. Sasser, Adele Clarke, Ruha Benjamin, Kim Tallbear, Michelle Murphy, Donna Haraway, Yu-Ling Huang, and Chia-Ling Wu, Michelle Murphy, "Forum on Making Kin, Not Population: Reconceiving Generations," *Feminist Studies* 45, no. 1 (2019): 168.

31. Population Matters, "An Equal World is a Better World," "Reproductive Health and Rights," in "Women's Rights," (June 13, 2021), https://populationmatters.org/womens-rights.

32. Graham Zabel, "Peak People: The Interrelationship Between Population Growth and Energy Resources," *Resilience* (April 2009). (June 13, 2021), www.resilience.org/stories/2009-04-20/peak-people-interrelationship-between-population-growth-and-energy-resources/.

33. See Luis I. Prádanos, *Postgrowth Imaginaries: New Ecologies and Counterhegemonic Culture in Post-2008 Spain* (Liverpool, UK: Liverpool University Press, 2018) and Ashish Kothari, Ariel Salleh, Arturo Escobar, Federico Demaria, Alberto Acosta, eds., *Pluriverse: A Post-Development Dictionary* (Chennai, India: Tulika Books, 2019).

34. Zoe Todd, "Indigenizing the Anthropocene," in *Art in the Anthropocene: Encounters Among Aesthetics, Politics, Environments and Epistemologies*, ed. Heather Davis and Etienne Turpin (London: Open Humanities Press, 2015), 244.

35. Heather Davis and Zoe Todd, "On the Importance of a Date, or, Decolonizing the Anthropocene," *ACME* 16, no. 4 (2017): 770.

36. Moore, *Anthropocene or Capitalocene?* 81.

37. Eileen Crist, "On the Poverty of Our Nomenclature," *Environmental Humanities* 3 (May 2013): 141, https://doi.org/10.1215/22011919-3611266.

38. Crist, "On the Poverty of Our Nomenclature," 141.

39. Sylvia Wynter, "Unparalleled Catastrophe for Our Species? Or, to Give Humanness a Different Future: Conversations," in *Sylvia Wynter: On Being Human as Praxis*, ed. Katherine McKittrick (Durham: Duke University Press, 2015), 22.

40. Wynter, "Unparalleled Catastrophe," 22.

41. Wynter, "Unparalleled Catastrophe," 24.

42. Stefania Barca, *Forces of Reproduction* (Cambridge: Cambridge University Press, 2020), 4.

43. Barca, *Forces of Reproduction*, 52, 1.

44. Barca, *Forces of Reproduction*, 5–7.

45. Kathryn Yusoff, *A Billion Black Anthropocenes or None* (Minneapolis: University of Minnesota Press, 2018), 51.

46. Yusoff, *Billion Black*, 62.

47. Kathryn Yusoff, *Billion Black*, xiii.

48. Kyle Powys Whyte, "Our Ancestors' Dystopian Now: Indigenous Conservation and the Anthropocene," in *Routledge Companion to the Environmental Humanities*,

ed. Ursula Heise, Jon Christensen, Michelle Niemann (London: Rougledge, 2017), 208–209.

49. Kim TallBear, "A Sharpening of the Already Present: Settler Apocalypse 2020," Department of Political Science, University of Alberta, YouTube, October 9, 2020, https://www.youtube.com/watch?v=eO14od9mlTA.

50. See Marisol de la Cadena, "Indigenous Cosmopolitics in the Andes: Conceptual Reflections Beyond Politics," *Cultural Anthropology* 25, no. 2 (May 2010): 334–370; Donna Haraway, *The Companion Species Manifesto: Dogs, People and Significant Otherness* (Chicago: Prickly Paradigm Press, 2003); Linda Hogan, *Dwellings: A Spiritual History of the Living World* (New York: Norton, 2007); Robin Wall Kimmerer, *Braiding Sweetgrass: Indigenous Wisdom, Scientific Knowledge and the Teachings of Plants* (Minneapolis: Milkweed Editions, 2015); Eduardo Kohn, *How Forests Think: Toward an Anthropology Beyond the Human* (Berkeley: University of California Press, 2013); Leah Penniman, *Farming While Black: Soul Fire Farm's Practical Guide to Liberation on the Land* (White River Junction, VT: Chelsea Green, 2018); Leanne Betasamosake Simpson, *Dancing on Our Turtle's Back: Stories of Nishnaabeg Re-Creation, Resurgence and a New Emergence* (Winnipeg, CA: Arp Books, 2011); Eduardo Viveiros de Castro, "Cosmological Deixis and Amerindian Perspectivism," *The Journal of the Royal Anthropological Institute* 4, no. 3 (September 1998): 469–488.

51. Kim TallBear and Angela Willey, "Critical Relationality: Queer, Indigenous, and Multispecies Belonging Beyond Settler Sex & Nature," *Imaginations* 10, no. 1 (July 25, 2019): 9, http://dx.doi.org/10.17742/IMAGE.CR.10.1.1.

52. Food and Agriculture Organization of the United Nations, "Productivity and Efficiency Measurement in Agriculture: Literature Review and Gaps Analysis," (February 2017): 17 (June 16, 2021), http://www.fao.org/3/ca6428en/ca6428en.pdf; Verena Seufert, Navin Ramankutty and Jonathan A. Foley, "Comparing the Yields of Organic and Conventional Agriculture," *Nature* 485 (2012): 229, https://doi.org/10.1038/nature11069;

53. See Jane Mt. Pleasant, "The Paradox of Plows and Productivity: An Agronomic Comparison of Cereal Grain Production under Iroquois Hoe Culture and European Plow Culture in the Seventeenth and Eighteenth Centuries," *Agricultural History* 85, no. 4 (2011): 460; Miguel A. Altieri, "Linking Ecologists and Traditional Farmers in the Search for Sustainable Agriculture," *Frontiers in Ecology and the Environment* 2, no. 1 (February 2004): 37.

54. TallBear and Willey, "Critical Relationality," 2.

55. Matthew Balley et al., "Which Racial/Ethnic Groups Care Most about Climate Change?" *Climate Change Communication* (April 16, 2020), https://climatecommunication.yale.edu/publications/race-and-climate-change/.

56. M. NourbeSe Philip, *Zong!* (Middletown, CT: Wesleyan University Press, 2011), 2.

57. Édouard Glissant, *The Collected Poems of Édouard Glissant*, ed. Jeff Humphries and Jane Humphries, trans. Melissa Manolas (Minneapolis: University of Minnesota Press, 2005), 212–213.

BIBLIOGRAPHY

Altieri, Miguel A. "Linking Ecologists and Traditional Farmers in the Search for Sustainable Agriculture." *Frontiers in Ecology and the Environment* 2, no. 1 (February 2004): 35–42.

Baker, Joelle, Paulla Ebron, Rosa Ficek, Karen Ho, Renya Ramirez, Zoe Todd, Anna Tsing, et al. "The Snarled Lines of Justice. When Women Warriors Map a New History of the Anthropocene." *Orion Magazine* (November 20, 2020), https://orionmagazine.org/article/the-snarled-lines-of-justice/?mc_cid=bf6614eb60&mc_eid=cdbabfa945.

Balley, Matthew, Edward Maibach, John Kotcher, Parrish Bergquist, Seth Rosenthal, Jennifer Marlon and Anthony Leiserowitz. "Which Racial/Ethnic Groups Care Most about Climate Change?" *Climate Change Communication*, April 16, 2020, https://climatecommunication.yale.edu/publications/race-and-climate-change/.

Barca, Stefania. *Forces of Reproduction*. Cambridge: Cambridge University Press, 2020.

Bignall, Simone and Daryle Rigney. "Indigeneity, Posthumanism and Nomad Thought: Transforming Colonial Ecologies." In *Posthuman Ecologies: Complexity and Process after Deleuze*, edited by Rosi Braidotti and Simone Bignall, 159–181. New York: Rowman & Littlefield, 2019.

Blaser, Mario, and Marisol de la Cadena. "Introduction: Pluriverse Proposals for a World of Many Worlds." In *A World of Many Worlds*, edited by Marisol de la Cadena and Mario Blaser, 1–22. Durham: Duke University Press, 2018.

Bradshaw, Corey A.J., Paul R. Ehrlich, Andrew Beattie, Gerardo Ceballos, Eileen Crist, Joan Diamond, et al. "Understanding the Challenges of Avoiding a Ghastly Future." *Frontiers in Conservation Science* 1 (January 2021). https://doi.org/10.3389/fcosc.2020.615419.

Brondizio, E. S., J. Settele, S. Diaz and H. T. Ngo. "Global Assessment Report on Biodiversity and Ecosystem Services of the Intergovernmental Science-Policy Platform on Biodiversity and Ecosystem Services." IPBES (2019). https://www.ipbes.net/global-assessment.

Ceballos, Gerardo, Paul R. Ehrlich, and Rodolfo Dirzo. "Biological Annihilation Via the Ongoing Sixth Mass Extinction Signaled by Vertebrate Population Losses and Declines." *PNAS* 114, no. 30 (2017): 6089–6096.

Crist, Eileen. "On the Poverty of Our Nomenclature." *Environmental Humanities* 3 (May 2013): 129–147.

Davis, Heather, and Zoe Todd. "On the Importance of a Date, or, Decolonizing the Anthropocene." *ACME* 16, no. 4 (2017): 761–780.

Dawson, Ashley. "Biocapitalism and De-extinction." In *After Extinction*, edited by Richard Grusin, 173–200. Minneapolis: University of Minnesota Press, 2018.

Elhacham, Emily, Liad Ben-Uri, Jonathan Grozovski, Yinon M. Bar-On and Ron Milo. "Global Human-Made Mass Exceeds All Living Biomass." *Nature* 588 (December 2020): 442–444. https://doi.org/10.1038/s41586-020-3010-5.

Escobar, Arturo. *Designs for the Pluriverse: Radical Interdependence, Autonomy, and the Making of Worlds*. Durham: Duke University Press, 2018.

Federici, Silvia. *Witches, Witch-Hunting, and Wome*n. Oakland: PM Press, 2018.

Glissant, Édouard. *The Collected Poems of Édouard Glissant*. Edited by Jeff Humphries and Jane Humphries, translated by Melissa Manolas. Minneapolis: University of Minnesota Press, 2005.

Grill, G., B. Lehner, and M. Thieme, et al. "Mapping the World's Free-Flowing Rivers". *Nature* 569 (2019): 215–221. https://doi.org/10.1038/s41586-019-1111-9.

Grooten, M. and R. E. A. Almond *Living Planet Report—2018: Aiming Higher*. WWF, 2018, Gland, Switzerland. https://wwf.ca/wp-content/uploads/2020/03/lpr 2018_summary_report_pages.pdf.

Kothari, Ashish, Ariel Salleh, Arturo Escobar, Federico Demaria, Alberto Acosta, eds. *Pluriverse: A Post-Development Dictionary*. Chennai: Tulika Books, 2019.

Maldonado-Torres, Nelson. "Thinking through the Decolonial Turn: Post-Continental Interventions in Theory, Philosophy, and Critique—An Introduction." *TRANSMODERNITY: Journal of Peripheral Cultural Production of the Luso-Hispanic World* 1, no. 2 (Fall 2011): 1–15.

McBrien, Austin. "Accumulating Extinction: Planetary Catastrophism in the Necrocene." In *Anthropocene or Capitalocene?* edited by Jason W. Moore, 116–137. Oakland: PM Press, 2016.

McNally, David. *Monsters of the Market: Zombies, Vampires and Global Capitalism*. Boston: Brill, 2011.

Mignolo, Walter. "Epistemic Disobedience, Independent Thought and Decolonial Freedom." *Theory Culture & Society* 26, no. 7–8 (2009): 1–23.

Moore, W. Jason, ed. *Anthropocene or Capitalocene? Nature, History, and the Crisis of Capitalism*. Oakland: PM Press, 2016.

Nixon, Rob. *Slow Violence and the Environmentalism of the Poor*. Cambridge: Harvard University Press, 2011.

Patel, Raj and Jason W. Moore. *A History of the World in Seven Cheap Things: A Guide to Capitalism, Nature, and the Future of the Planet*. Berkeley: University of California Press, 2017.

Perry, Imani. *Vexy Thing: On Gender and Liberation*. Durham: Duke University Press, 2018.

Philip, M. NourbeSe. *Zong!* Middletown, CT: Wesleyan University Press, 2011.

Plumwood, Val. "The Crisis of Reason, the Rationalist Market, and Global Ecology." *Millennium: Journal of International Studies* 27, no. 4 (1998): 903–925.

Plumwood. *Environmental Culture: The Ecological Crisis of Reason*. New York: Routledge, 2002.

Prádanos, Luis I. *Postgrowth Imaginaries: New Ecologies and Counterhegemonic Culture in Post-2008 Spain*. Liverpool, UK: Liverpool University Press, 2018.

Red Nation, The. *The Red Deal: Indigenous Action to Save Our Earth*. Brooklyn: Common Notions, 2021.

Shiva, Vandana. *Earth Democracy*. Cambridge: South End Press, 2005.

Smith, Mick. *Against Ecological Sovereignty: Ethics, Biopolitics, and Saving the Natural World*. Minnesota: University of Minnesota Press, 2011.

Strathern, Marilyn. Jade S. Sasser, Adele Clarke, Ruha Benjamin, Kim Tallbear, Michelle Murphy, Donna Haraway, Yu-Ling Huang, and Chia-Ling Wu. "Forum on Making Kin Not Population: Reconceiving Generations." *Feminist Studies* 45, no. 1 (2019): 159–172.

Strona, Giovanni and Corey J.A. Bradshaw. "Co-extinctions Annihilate Planetary Life During Extreme Environmental Change." *Scientific Reports* 8, no. 16724 (2018): https://doi.org/10.1038/s41598-018-35068-1.

TallBear, Kim. "A Sharpening of the Already Present: Settler Apocalypse 2020." Department of Political Science, University of Alberta, YouTube, October 9, 2020. https://www.youtube.com/watch?v=eO14od9mlTA.

TallBear, Kim and Angela Willey. "Critical Relationality: Queer, Indigenous, and Multispecies Belonging Beyond Settler Sex & Nature." *Imaginations* 10, no. 1 (July 25, 2019). http://dx.doi.org/10.17742/IMAGE.CR.10.1.1.

Theriault, Noah and Audra Mitchell. "Extinction." In *Anthropocene Unseen: A Lexicon*, edited by Cymene Howe and Anand Pandian, 177–182. Punctum Books, 2020.

Todd, Zoe. "Indigenizing the Anthropocene." In *Art in the Anthropocene: Encounters Among Aesthetics, Politics, Environments and Epistemologies*, edited by Heather Davis and Etienne Turpin, 241–254. London: Open Humanities Press, 2015.

Tuck, Eve and K. Wayne Yang. "Decolonization is Not a Metaphor." *Decolonization: Indigeneity, Education & Society* 1, no. 1 (2012): 1–40.

Yusoff, Kathryn. *A Billion Black Anthropocenes or None*. Minneapolis: The University of Minnesota Press, 2018.

Yusoff, Kathryn. "The Inhumanities." *Annals of the American Association of Geographers* 111, no. 3 (2021): 663–676.

Whyte, Kyle Powys. "Our Ancestors' Dystopian Now: Indigenous Conservation and the Anthropocene." In *Routledge Companion to the Environmental Humanities*, edited by Ursula Heise, Jon Christensen, Michelle Niemann, 206–215. London: Routledge, 2017.

Wynter, Sylvia. "Unparalleled Catastrophe for Our Species? Or, to Give Humanness a Different Future: Conversations." In *Sylvia Wynter: On Being Human as Praxis*, edited by Katherine McKittrick, 9–89. Durham: Duke University Press, 2015.

Zabel, Graham. "Peak People: The Interrelationship Between Population Growth and Energy Resources." *Resilience* (April 2009). www.resilience.org/stories/2009-04-20/peak-people-interrelationship-between-population-growth-and-energy-resources/.

Chapter 1

Decolonize, ReIndigenize

Planetary Crisis, Biocultural Diversity, Indigenous Resurgence, and Land Rematriation

Leonardo E. Figueroa Helland, Abigail Perez Aguilera, and Felix Mantz

INTRODUCTION

Across Mother Earth, Indigenous peoples have nurtured intimate relations with their lands, enabling the coevolution of locally attuned land governance and management matrices where biological and cultural diversity are reciprocally enriched. Indigenous peoples are thus keystone societies, fostering ecosystem richness and regeneration. The destruction of Indigenous cultures, land dispossession, and erosion of Indigenous self-determination, knowledges, and cosmovisions undermine sustainability. Colonialism—in its multiple forms—underpins Euro-modernity, Western "civilization," capitalism, statecraft, and other power structures eroding Indigeneity and biocultural diversity. Mother Earth's ecological crisis is rooted in these injustices. Conversely, the resurgence of Indigenous peoples and lifeways, particularly through the defense, rematriation, and restitution of Indigenous land bases and the reclamation of communal and non-anthropocentric land governance based on Indigenous knowledges (IKs) and sovereignty is key to the cyclical regeneration of bountiful life. Defending and revitalizing Indigeneity while dismantling colonial and associated power structures are indispensable decolonizing requirements to overcome and p/refigure alternatives to the systems driving "Anthropocene" crises. There can be no healing to humanity's catastrophic rift with Mother Earth without Indigenous resurgence and decolonization.

This chapter articulates the above arguments as follows. (I) We first examine the "Anthropocene" as a colonial process rooted in the systematic erasure of Indigeneity and biocultural diversity. (II) We then examine how Indigenous peoples are keystone biocultures central to regenerating biodiversity and addressing climate change. (III) Thereafter, we explore Indigenous biocultures as nurturers/curators of territories of life. (IV) We then explore the keystone biocultural role of Indigenous peoples in (a) East/Northeast Turtle Island and (b) Mesoamerican Abya-Yala. (V) Afterward, we critically engage the necropolitical violence of, and resistance to, the genocide-ethnocide-ecocide nexus undermining Indigenous peoples and biocultures. (VI) In lieu of a conclusion, we argue for decolonization and re-Indigenization, not inclusion, particularly vis-à-vis moves toward the "inclusion" of IKs and biocultural diversity in global environmental/climate policy, which signal colonial equivocations that sidestep the frontal politics of Indigenous decolonization, resurgence, and land rematriation.

THE ANTHROPOCENE'S COLONIAL ROOTS: INDIGENEITY AND BIOCULTURAL DIVERSITY UNDERMINED

We live in the so-called Anthropocene—a geological epoch of converging crises, including climate change and biodiversity loss/species extinction.[1] While this term suggests that "humanity" has become the dominant driver of catastrophic planetary disruptions, not all humans are responsible.[2] The socially undifferentiated "Anthropos" obscures injustices, inequalities, and power, propagating a universalizing discourse that misallocates responsibility to humanity-as-a-whole while failing to confront the roots of eco-catastrophes.[3] Not caused by undifferentiated and universal humanity, eco-catastrophes rather result from the historical development of Western "civilization" and Euro-"modernity." Humans require an elementary metabolic interchange with nature to satisfy their needs, approximately 800 kg per person annually. This is, on average, approximately tenfold higher in industrial societies—40 to 70 *tons* per person annually.[4] Diverse systems, classes, races, cultures, genders, and peoples organize metabolic matter/energy flows differently, engendering—sometimes dramatically—contrasting relations with Mother Earth. Moreover,

> [the] causes and consequences, benefits and costs [of eco-crises], are sharply uneven. . . . The unpaid or barely paid work of denigrated classes . . . –women, enslaved . . . , colonized . . . , impoverished people—has . . . enrich[ed] states, capitalists, whites, and men, in extractive and polluting industries. . . .[5] These

same classes bear . . . exposure to the ecological harms of the Anthropocene while carrying little responsibility and gaining marginal to no benefit.[6]

Challenging Anthropocene equivocations, Indigenous, Black, decolonizing, eco-feminists, eco-Marxist, posthumanist, and queer voices underline the anthropocentric, (hetero)patriarchal, Eurocentric, colonial, racial, and capitalist roots of global crises, calling this epoch Androcene, Eurocene, Plantationocene, Homogecene, and Capitalocene.[7] At root is not the abstract "Human" nor "some shared human . . . inclination towards destruction,"[8] but an "imperial mode of living" (IML)[9] spread by a modern/colonial Western(izing) civilization. This is the embodiment of the voracious matter-energy metabolism of an extractivist, industrialist, and developmentalist model that has altered Mother Earth's geological trajectory. This model is one with the history of colonialism and imperialism, the capitalist world-system, unequal ecological exchange, exploitation, and underdevelopment. Simply, "[w]ithout . . . empire, the industrial revolution would have been physically impossible"; the rich countries' "prosperity" is built by monopolizing the benefits from Earth and externalizing environmental harms.[10]

Western "modernity" relies on the colonial continuum of massive socio-environmental and climate debts fueled by the suffering of billions of humans and untold nonhumans "who over . . . generations have borne . . . injustices unquantifiable, both social and ecological."[11] Unjust legacies continue through racially uneven geographies of vulnerability and death from the Anthropocene crises.[12] Central is the ongoing destruction of Indigenous peoples for over five centuries, constituting the Anthropocene's "founding" violence.[13] The invasion and colonization of Indigenous Abya-Yala/Turtle Island ("the Americas"), alongside the kidnapping and enslavement of Indigenous Africans, unleashed the monocultural homogenization of human and nonhuman life, for example, in *encomiendas*, *reducciones* (reductions), plantations, reservations and boarding schools, devastating peoples and environments, hence the "Plantationocene."[14] In Abya-Yala/Turtle Island colonial conquest wars, genocide, forced labor, and diseases brought by Europeans caused a roughly 90% decline of Indigenous populations, from 54 million to 61 million (c. 1492) to about six million (c. 1650).[15] The racially charged imperial theft of whole continents alongside the devastation and enslavement of Indigenous peoples initiated the Eurocentric reorganization of the planet into a world-capitalist monoculture of extraction, accumulation, and waste whose unbridled expansion is breaching planetary boundaries.[16] The genocide of Abya-Yala's Indigenous peoples and the destruction of their land tending, food systems, and social economies were of geo-historical proportions, triggering a radical drop in atmospheric CO_2 between the 1500s and 1610. This "Orbis Dip," suspected of contributing to the Little Ice Age, marked the onset of a rapidly consolidating greenhouse-gas-intensive Eurocentric world-capitalist economy.[17] A "Capitalocene"[18] epoch is only possible because of coloniality and racism, "as proto-capitalists, colonists, and Christians all drew on white supremacy . . . [in] severing Indigenous peoples from their land and labor."[19]

The breaching of Mother Earth's so-called planetary boundaries that has put us on the precipice of "Hothouse Earth"[20] results from this colonial continuum that still undermines the biocultural diversity embodied by Indigenous peoples. The Anthropocene embodies centuries-old logics of dispossession, eco-genocide, and the destruction of Indigenous peoples based on doctrines of "discovery" and dominion, and patriarchal epistemologies of mastery over a feminized nature.[21] Euro-modernity and Western civilization are built on the destruction of Indigenous societies with healthy communal socioecological relations that have nurtured biocultural diversity and ecosystem regeneration for millennia.[22] For over 500 years, the destruction of up to 5,000 of the approximately 12,000 ethnolinguistic cultures that existed before European colonialism has been carried out in the name of Western modernity and civilization.[23] Most such cultures are Indigenous and non-Western, consisting of intimate regenerative relations with ancestral land bases.[24]

Despite colonial genocidal/ecocidal/ethnocidal violence and ongoing erasure, there are approximately 400 million Indigenous people worldwide, constituting over 5,000 nations and accounting for 5–10% of the global population.[25] These populations comprise 4,000 of the world's 7,000 remaining languages, embodying the widest ethnic/linguistic/cultural diversity—90% according to First Peoples Worldwide.[26] Yet statistics about Indigenous peoples underestimate and perpetuate erasure since they rarely account for the colonial continuum of historical, intergenerational, as well as ongoing processes and power structures continuously producing de-Indigenization.[27] Statistics obscure how colonization, de-Indigenization, and assimilation are not one-time events, but social destructuration and dismemberment processes resulting from institutionalized power structures undergirding modern and settler apparatuses through the normalized erasure of Indigenous identities, (bio)cultures, territorialities, and land relations. The destruction and dispossession of Indigenous peoples and lands result not only from "fast/manifest" violence, but also "slow-onset" eco/ethno/genocide by institutions of empire, the state, industrial capitalism, the market, and associated apparatuses, including education, religion, culture, and the law.[28] Given ongoing histories of coloniality and racism, including external and internalized oppression, many who would otherwise identify with and practice Indigenous lifeways are compelled to leave behind communal identities, languages, knowledges, and land-based lifeways under colonizing and westernizing pressures. Often Indigenous identity erasure is intergenerationally transmitted as a—sometimes strategic, sometimes internalized—survival response within oppressive contexts of coloniality. Descendants can grow reluctant to maintain, display, or transmit Indigenous identity, language, knowledge, or lifeways intergenerationally for fear of the high—sometimes lethal—social costs. Colonizing and "modernizing" pressures, including national/state language standardization, educational assimilation/acculturation, evangelization, epistemological Eurocentrism, and globalization's monocultures contribute to projections that 50–90% of the world's languages will be erased by 2100.[29] Language diversity co-constitutes cultural and epistemological diversity; hence, Indigenous language erasure facilitates epistemicides—the erasure of IKs.[30]

By destroying Indigenous diversity, colonization and assimilation enable dramatic biodiversity loss. Indigenous cultures have intimately coevolved for millennia with unique habitats, actively co-nurturing their biodiversity through relationally tailored communal knowledge-praxes. For instance, the Maya have inhabited Mesoamerica for at least 8,000 years,[31] and the so-called Pygmies (including Baka, Aka, Twa, Mbuti peoples) have lived in African rainforests for at least 60,000 years.[32] Each Indigenous culture has nurtured bioculturally adapted land-based knowledges, practices, and social-ecological organizations. Rooted in cosmovisions that sacralize non-anthropocentric communality, such knowledges buttress social ecologies dedicated to the cyclical regeneration of the commons and the deliberate fostering of complementary diversity—not extraction, exploitation, or accumulation. Globally, IKs have enabled the estimated nurturance of at least 1,400 new species and countless varieties enhancing agro/biodiversity. Through (bio)mimicking Mother Earth's biodiversity and landscape heterogeneity, Indigenous peoples have diversified into thousands of cultures, each adapted to unique landscapes. Today the greatest biodiversity exists alongside the greatest cultural/linguistic diversity. For example, New Guinea alone has 800 languages, and Oaxaca (Mexico) 150 in only 92,000 km^2, including Mesoamerica's highest biodiversity area—the Zoque Rainforest.[33]

Eroding Indigenous diversity undermines locally tailored land governance, and consequently biodiversity. Estimates show rapid biodiversity losses over the last centuries, thereby compromising ecosystem function.[34] Recent decades of neoliberal globalization—modernity/coloniality's latest stage—have accelerated interlocked trends in biocultural diversity loss. The Index of Linguistic Diversity and the Living Planet Index (LPI)[35] of species populations report highly correlated declines of at least 30% since the 1970s. Global linguistic diversity has declined by 20%. Abya-Yala/Turtle Island faces an alarming 60% linguistic diversity loss, with 75% in its Neotropics, while declines for Oceania, Africa, the Indo-Pacific, and Palearctic are 30%, 20%, 30%, and 30%, respectively.[36] The linguistic imperialism and linguicide (systematic erasure of languages) started by colonization continues,[37] facilitating and correlating with biodiversity loss. The most dramatic species declines are in the Indo-Pacific (65%), Abya-Yala's Neotropics (50%), and the Afrotropics (40%).

Indigenous lifeways nurtured over millennia of intimate co-evolving relations with specific places and ecosystems have enabled long-living non-anthropocentric communities and relational cosmologies, embedding human lifeways into the co-creative cyclical regeneration of life-webs. They have thus sustained and enhanced biodiversity and "carbon sinks" by engendering a co-dependence of biological and cultural diversity. Contrastingly, colonial, Eurocentric, patriarchal, and capitalist industrial civilization has—in a few hundred years—pushed the planet to mass extinction, biocultural obliteration, and climate catastrophe. The last five centuries have seen biocultural diversity supplanted with anthropocentric monocultures of extraction, production, accumulation, consumption, and waste that undermine socioecological reproduction and resiliency while risking biospheric collapse and—potentially irreversible and catastrophic—Earth system destabilization.[38] The "Homogocene"[39]—this epoch of tragic, often forcible homogenization and reduction of human-and-ecological diversity—embodies the "Anthropocene's" "foundational" violence, which recursively reenacts the brutal colonial logic that systematically imposes Euro-modernity through geno/ethno/ecocide, forced integration, climate change, and radical biospheric simplification.[40] Colonizers have refused to see, respect, or learn from the diverse landscapes, climates, flora and fauna, and the millennial cultures nurturing them.

Addressing the colonial roots of the Anthropocene crises indispensably requires decolonization and re-Indigenization. As *The Red Deal* states: "There is no hope for restoring the planet's fragile and dying ecosystems without Indigenous liberation and land rematriation. . . . The choice is: decolonization or extinction."[41] Rematriation, Lenape/Shawnee scholar Steve Newcomb explains, requires restoring people to their rightful place in sacred relationship with their ancestral land and Mother Earth. This involves, inter alia, returning lands to their original Indigenous caretakers; enabling the resurgence of Indigenous identities, cultures, and lifeways; restoring Indigenous communal land governance, management, and commons tenure; defending and expanding Indigenous self-determination, rights, and sovereignty; and reclaiming and revitalizing IKs, cosmovisions, languages,[42] and relational conviviality worldwide. It is through the resurgence of Indigenous communality, reciprocity, and land-based cosmovisions, including projects for the defense, sovereignty, and rematriation of Indigenous lands, that the creative reconstitution of biocultural territories of life can form the basis of struggles for worlds other than—and outside of—the death-worlds and crises of colonial, capitalist, extractivist, and patriarchal systems.

Today we face two opposing pathways. One takes us down to the dominant market-based, state-centric, and intergovernmentally sponsored schemes premised on false solutions and dangerous techno-fixes to climate, biodiversity, food, health, migration, and political crises. These include a neoliberal capitalist green economy that commodifies nature through payments for ecosystems services (PES), carbon/biodiversity pricing, markets and offsetting, geoengineering and bioengineering, so-called climate-smart agriculture, and allegedly "green" energy industrial-scale mega-development projects. These deepen and expand coercive extractivism, land, water, and green grabbing, forced displacement, neofascist migration "management," "population control," and border imperialism, securitization, and militarization.[43] They re-entrench the anthropocentric, colonial, patriarchal, capitalist, neoliberal, and militarizing frameworks at the root of epochal crises. *The Red Deal* warns, "[w]hat creates crisis cannot solve it;" "colonialism . . . and capitalism destroyed this world . . ." and "must be overturned for this planet to be habitable for human and other-than-human relatives to live dignified lives."[44] Beyond the hubris of a civilization bent on mastering Mother Earth—including human and nonhuman "others"— Indigenous, decolonial, Afro-Indigenous, Black abolitionist, peasant, Global South, feminist and QTBIPOC subaltern, frontline and grassroots approaches move us beyond the obsession with managerial control and the "resilience" of violent systems producing syndemic[45] crises. They shift us to restoring and nurturing regenerative, complementary, reciprocal, and bioculturally diverse territories of life based on care for land, human, and other-than-human communities, and honoring Mother Earth.[46]

BIOCULTURAL DIVERSITY AND CLIMATE CHANGE: INDIGENOUS PEOPLES AS KEYSTONE CULTURES

Biocultural diversity is life's diversity in all its biotic and human manifestations, interwoven and co-evolving as continuing outcomes of complex, socio-ecologically adaptive systems.[47] It includes the diversity of plants, animals, habitats, ecosystems, landscapes, human cultures, languages, and knowledges. Not merely parallel, these diversities embody a complex whole: the living, interlocked cumulative effects of co-adaptation, co-influence, and co-design within integrated socioecological metabolisms. For Indigenous peoples, interpersonal relational intimacy weaves humans and territories into integrated communities where holistic life regeneration depends on reciprocal human–nonhuman nurturance. Biocultural coevolution characterizes Indigenous societies that have dwelled, nurtured, and been nurtured by their landscapes long before colonization. Biodiversity depends on Indigenous resiliency.[48] Around 80% of global biodiversity occurs in the 20–25% of the planet's terrestrial surface still in Indigenous management/tenure.[49] Indigenous communities live in areas containing most of the world's genetic resources and biodiversity hotspots.[50] Biodiversity hotspots and high biodiversity "wilderness" areas encompass 70% of language diversity, primarily Indigenous languages embodying millennia of land-based knowledges and cosmovisions—facing erasure.[51] Thus, most biodiversity remains only due to the keystone role of Indigenous cultures in nurturing and defending biotic communities as sacred, especially vis-à-vis (ongoing) colonization, capital accumulation, state developmentalism, and extractivism. Hence, protecting and fostering biodiversity depends on defending, restoring, and revitalizing Indigenous cultures, and vice versa—that is, the biocultural axiom.[52]

Indigenous resurgence, including defense and restoration of community lands, is key to addressing climate change.[53] For instance, through community forest management (CFM), Indigenous peoples have safely and effectively managed carbon cycles, regenerating carbon sinks while protecting biodiversity for millennia without risky techno-fixes, conservation enclosures, or market mechanisms.[54] About 25% of tropical/subtropical forest carbon is in territories collectively managed by Indigenous and local communities—although one-third lack formally recognized tenure rights.[55] The territories tended by customary Indigenous and local communities are underestimated given limited tenure recognition. While some quantify that Indigenous peoples manage or have tenure rights over a quarter of the world's land surface,[56] others specify that Indigenous and local customary tenure and claims likely cover over 50% of Earth's land (about 6 billion hectares), though "legal" title covers only 10–15% and formal usufruct and/or management rights another 10%.[57] Lack of recognition allows governments—and corporations licensed by them—to label communities "illegal" in their own territories, enabling land grabs.[58] About 80% of remaining forests worldwide are protected by 370 million Indigenous people whose sovereign territories are threatened by miners, ranchers, plantation owners, and their government allies. As Indigenous peoples are displaced, so are their intimate biocultures and knowledges.[59] The *Local Biodiversity Outlooks 2* (2020) warns "the lands of Indigenous peoples are becoming islands of biological and cultural diversity surrounded by areas in which nature has . . . deteriorated" (39).

Indigenous communities live in or adjacent to about 85% of protected areas, and Indigenous-managed forests show slower or no degradation and deforestation.[60] Securing Indigenous collective tenure and strong customary community governance, and restoring it where it has been lost, brings lower or no deforestation and soil degradation. It sustains ecosystems and restores damaged habitats. Even when compared to government protected areas, Indigenous tenure shows better protection or even enhancement of biodiversity, climate change mitigation/adaptation capacity, resiliency, and ecosystem functions, with Indigenous CFM consistently showing increased forest carbon stocks.[61] Almost 38 billion tons of carbon are in the living biomass of community forests worldwide.[62] Such community-based forestry is still practiced in roughly 28% of countries' forests.[63] Support for Indigenous communities and management systems is key to overcoming Anthropocene crises and regenerating ecosystems.[64] However, they should not have to carry the burden of offsetting, compensating, or resolving (e.g., via carbon/biodiversity market mechanisms like REDD+) damages caused by others' imperial mode of living.[65] Indigenous communities/practices can thrive only if root crises drivers (e.g., capitalism, colonialism, industrial/consumer civilization, patriarchy, extractivism) are mitigated/dismantled at the source.[66]

CURATING TERRITORIES OF LIFE

As keystone cultures, Indigenous peoples nurture, restore, and enrich ecosystem (agro)biodiversity, soil fertility, and carbon regulation capacity through knowledges, cosmovisions, and communal socioecological governance rooted in intimate land relations.[67] Indigenous management fosters rich biodiverse land/water/seascapes by working with, mimicking, and enhancing species complementarity, relational ecosystem functions, and regenerative cycles, often producing gardened landscapes.[68] Long-term dwelling, coevolution, and cultural respect that listens to the land underpin accurately adapted and sensitive knowledges embodying a communal ethic where life's cyclical renewal for indefinite (re)generations depends on reciprocal human–nonhuman communities.

Often, major ecosystems, like forests, erroneously assumed—Eurocentrically—as "untouched," "pristine," or "wild" have been curated through millennia of Indigenous socioecological co-design with the land.[69] Indigenous cosmovisions promote customary commoning practices underpinning regenerative economies and sustainable food systems. These embody ecologically attuned and ritually regulated calendars for resource harvesting, and complex spatially mixed designs of forest gardens, sacred groves, and buffer zone maintenance. To illustrate, many Indigenous cultures actively foster and induce the regeneration and enrichment of ecological succession cycles.[70] Such shifting cyclical forest management has had agroecological and agrobiodiversity-enhancing effects, providing decolonizing pathways to ecosystem restoration.[71] Diverse Indigenous cultures practice(d) complex bioregionally and locally tailored forest gardening, orchards, sacred groves, and integrated cyclical ecosystem management. Consider Mesoamerica's Maya and Huastec cultures,[72] Amazonia's Bora, Amuesha, Kayapo, and Ka'apor cultures,[73] Turtle

Island's Haudenosaunee and Lenape,[74] East Africa's/Tanzania's WaChagga,[75] Borneo's Dayak,[76] and South Asian Adivasi.[77] Their forest management integrates woody perennials in spatial mixtures or temporal sequences with crops and/or animals. Indigenous agroforestry, sometimes including fire regimes like controlled burns, offers advantages over conventional agriculture and forestry.[78] They increase food crops, fodder, fruits, nuts, building materials, medicinal plants, and non-timber products. Especially when unimpeded by colonial/state/capitalist land encroachments, they provide socioeconomic benefits like food sovereignty, communal governance, self-determination, and reduced migration pressures. Ecologically, they regenerate soil fertility, foster biodiversity, and increase carbon sequestration in biomass and soils.[79]

Land-based cosmovisions underpin communal Indigenous land management, emphasizing the roles and responsibilities of humans to collectively contribute to and celebrate the cyclical renewal of the richness, bounty, and complementary diversity of the living community where all is related. Notions like *Utz'Kaslemal* (Maya), *Suma Qamaña* (Ayamara), *Suma Kawsay* (Quechua), *Ubuntu/Ukama* (Bantu), *Nande Reko* (Guarani), *Bimaadaziwin* (Anishinaabe) embody variants of the "Indigenous paradigm"[80]: to always live collectively in ways that contribute to nurture the continuous rebirth of plentiful vitality for the whole community of life. These starkly oppose the anthropocentric and patriarchal ontologies of Abrahamic—particularly (though not exclusively) Christian—worldviews, and modernity's linear narratives of anthropocentric exceptionalism (whether theological, rational/logocentric, or evolutionary), progress, development, accumulation, and Man's civilizational superiority based on subalternizing, desacralizing, objectifying, dominating, commodifying, and mastering nature.[81] Consider now examples of Abya-Yala's/Turtle Island's keystone biocultures showing how Indigenous land governance promotes biocultural diversity and sustainability.

KEYSTONE BIOCULTURES ACROSS ABYA-YALA/TURTLE ISLAND

East/Northeast Turtle Island

In 1974, the Ganienkeh Council Fire—an ongoing project for an independent/sovereign Kanien'kehá:ka (Mohawk) traditional community on rematriated ancestral land—manifested:

> What is regarded as progress . . . is a road to destruction. . . . The result has been . . . pollution of air, land, water and the human mind. . . . [A]bused nature repays in kind. . . . [T]he profits . . . of a few tycoons and the worship of . . . technology . . . brought the world to the brink of destruction.[82]

Danny Beaton (Kanien'kehá:ka/Mohawk) and John Bacher illustrate how for millennia the Haudenosaunee Confederacy "developed a civilization [of] remarkable sustainability with the life-force of the blessed area around the Finger Lakes and . . . Mohawk Valley" that was based on "a gentle way of life" protecting Mother Earth's vitality. While cultivating "the three sisters of corn, beans and squash" and fishing sustainably, respecting reproductive cycles, the Haudenosaunee also "cared for the forests . . . [as] a great garden" through "gentle landscape modification," assisting the flourishing of "a variety of tree species carrying edible fruits and nuts," including "[h]ickories, butternut, black walnut, cherries, plums, and pawpaws."[83] Socioecological stability for these forest gardens was maintained through skilled land governance, relational diplomacy, and regulations like the Great Law of Peace. Anglo colonial/settler-colonial invasion devastated Haudenosaunee socioecologies.[84] Water bodies suffered abuse and fishing became nearly impossible without causing extinction. The vicious burning of forests, pillaging of communities (e.g., during the US Army's Sullivan Expedition), the opening of lands for settlers, monocultures, and livestock—altogether destroyed polycultures, forest gardens, ecologies, and peace; by the 1800s, settler-colonial "clear-cutting . . . of the last well-wooded lands in southwestern Ontario . . . [brought such] devastation . . . that . . . trees could not regenerate. Much of . . . Ontario [was] . . . turned into a desert of dangerously marching sands."[85]

Preceding this colonial violence, Southern Ontario had over 80% forest cover—now only 10% remains; this dramatic biodiversity destruction of "settler agriculture and urbanization removed perennial vegetation from 70% of an entire landscape in under three hundred years."[86] Indigenous peoples who still inhabit this land suffer from dramatic land loss, forced relocation, loss of biodiversity, polluted ecosystems, and multiple health impacts, while agriculture now heavily favors external inputs like pesticides and fertilizers.[87] The complex Haudenosaunee forest gardening system—devastated by Euro-American settler-colonialism—exemplifies the integrated ecosystem tending of Indigenous cultures who profoundly shaped biomes and culturally coevolved with Turtle Island landscapes. Across most Eastern/Northeastern Turtle Island, bioculturally diverse landscapes prevailed under Indigenous governance. Millennia of Indigenous landscape nurturance fostered expansive agroecological and forest orchard gardens, creating vast forested landscapes mistakenly deem(ed) "wild" and "untouched" by settler colonizers. Euro-American intrusion imposed alien settlement and land-uses that "radically altered the environment," massively displacing "the Indigenous inhabitants and ceasing their land-use habits."[88]

When Indigenous governance/sovereignty was unimpeded, this region's bioculturally diverse landscapes ensured ecosystem integrity and agroecological food sovereignty. Indigenous nurturance of vast forest garden/orchard landscapes had many social-environmental benefits, like maintaining a balanced carbon cycle (e.g., avoiding excess GHG by regenerating carbon sinks through forest polycultures), employing controlled fires and soil fertility management. By deliberately managing multistage ecological succession cycles, Indigenous Turtle Islanders of this region fostered over thirty-five herbaceous plants, twenty-one varieties of fruit and berry trees, and sixteen varieties of mast and nut trees (including numerous oak species).[89] They exemplify a culture acting "as keystone species by disproportionately manipulating the environment and component species relative to their . . . abundance."[90] The mast was critical for carbohydrates and fat calories and at least thirty tree species and genera were favored and used, plus twenty or more cultivated fruit and berry trees.[91] Indigenous management shaped tree diversity and "the entire historical development of . . . eastern oak and pine forests, savannas and tallgrass prairies."[92]

During much of the Holocene, Indigenous forest gardening prevailed across most of the Northeast, Appalachian, Southeast, and Great Lakes regions, maintained by sophisticated fire regimes and forest orchard gardening.[93] For example, oak-dominated cultivated forests once covered Lenape lands, which exemplify a forest gardening culture.[94] Around 7,000 to 6,000-year-old Minisink and Munsee Lenape fire practices created *moshulus*—open fields of rich blackened soils that enabled the planting of forest gardens and nut/fruit orchards, which also became game parks, by deliberately propagating preferred species.[95] Controlled burning was not homogeneous; a "burn index"[96] reveals a spectrum and matrix design of diverse yet complementary and rotating landscapes rich in bioculturally useful species, combining northern hardwood forests with landscapes of southern species, and others for polyculture agriculture, which were gradually restored to close canopy forest in long cycles of ecological restoration and enrichment.[97] Today, the dispossessed Lenape, who still struggle to reclaim sovereignty and lands,[98] see Lenapehoking territories and forest gardens overtaken by settler megalopolises, monocultures, and extractivism.

Complex controlled Indigenous fire helped regenerate soil fertility and was integrated into the management of ecological succession cycles to nurture richly diverse polycultures, thus constituting pyrogenic forested landscapes, often misunderstood by settlers as "wild/pristine."[99] Forests from Maine to Florida and west past the Mississippi were no wildernesses, but Native American cultivated gardens deliberately managed via periodical burning and selective tree girdling and planting.[100] This cultivated forest landscape of "fruitful gardens" counteracts the settler-imagined "pristine . . . wilderness" which probably "never existed"; landscapes were Indigenously co-designed with the land, providing "sustenance for a diverse community of life."[101] Elsewhere in Turtle Island, including California where climate change and poorly planned, excessive urbanization heighten wildfire vulnerability, some settler interest in Indigenous fire regimes has emerged, enabling political conditions supportive of Karuk and Yurok nations' fire reclamation/revitalization programs.[102] Beyond Abya-Yala/Turtle Island, Indigenous peoples like Aboriginal Australians[103] and East African Maasai pastoralists[104] have similarly employed fire regimes to regeneratively manage biodiversity-rich polycultural landscapes.

Western research only recently started recognizing IKs and practices as necessary for flourishing forests.[105] Agro-ecology/forestry and permaculture are slowly considering that regional Indigenous management/governance/cultures are keystones to sustainable and just life-supporting systems long predating settler science.[106] Still, decolonization and Indigenous resurgence demand going beyond inclusion and recirculation of specific Indigenous practices that converge with settler/colonial interests. Indigenous-led revitalization must also go beyond technical aspects of land management to encompass land, culture, knowledge, and sovereignty reclamation. In the northeast, notable contemporary projects include The Haudenosaunee-co-led Northeast Farmers of Color Land Trust (NEFOC-LT),[107] The Ganienkeh Council Fire,[108] Kanatsiohareke Mohawk Community,[109] the Three Sisters Sovereignty Project,[110] the Queer OnKwehowe Land Project led by Dioganhdih (Kanien'kehà:ka/Mohawk),[111] the Cayuga Share Farm (Gayogohó:nǫˀ),[112] Nibezun (Wabanaki),[113] Schaghticoke First Nations Land Reclamation Project,[114] and the Haudenosaunee Environmental Task Force,[115] among others.

Mesoamerican Abya-Yala

In 2018, the Zapatista Army of National Liberation issued a communique restating the central justification for Indigenous struggle:

> "The West" . . . self-designated symbol of civilization, goes out, destroys, depopulates, and then retreats and closes its doors, leaving big capital to continue . . . its business. . . . [O]riginary peoples . . . must struggle and defend . . . territories—to death if necessary . . ., there is no other choice.[116]

Originary ("Indigenous") Mesoamericans have rejected Eurocentric domination and rebelled against "acculturation" into modern states and capitalism because of and through *comunalidad:* a land-based non-anthropocentric communality based on collective reciprocity, mutualism, relationality, cyclicality, and consensual sovereignty and self-determination.[117] This buttresses resistance and resilience vis-à-vis the economic reductionism of colonial, Eurocentric, and capitalist civilization which, as Zapatistas underline, "wants to turn the whole world into plantation":

> As it develops, the system "discovers" new commodities . . . in . . . territories of . . . originary peoples: water, land, air, biodiversity, everything that hasn't already been . . . ruined lies in originary peoples' territories . . . , where the system . . . seeks (and conquers) new markets. [T]hese aren't just consumption markets . . . , the system seeks . . . to conquer territories and populations . . . to extract . . . whatever can be extracted, without regard for the wasteland left . . . The "development" and "progress" offered . . . [is] obtained via the death and destruction of populations and territories. That is how so-called "civilization" is founded.[118]

Furthermore,

> "Civilization" is what they . . . call . . . the destruction of our society and . . . culture, the massacres of the Indigenous, seizure of lands and wealth, humiliation . . . and contempt for our culture, mockery of our language, rejection of our clothing, disgust for our dark color . . . , the Earth's color [replicating Her mistreatment]. Now the same war against us has another name . . . , "modernization." . . . It's the voice of the same . . . who yesterday used whip and sword to conquer our land and today uses modernization to do away with us . . . it's the voice of the one who cannot conceive of any way of living other than at the cost of our deaths. It's the voice of the one who says that . . . Indigenous peoples will make progress only when they cease being Indigenous. . . . As our ancestors resisted wars of conquest and . . . extermination, we continue to resist.[119]

Originary Mesoamericans have opposed the desacralization, objectification, commodification, and exploitation of Mother Earth underpinning capitalist accumulation, property, profit, and extractivist statecraft. They have resisted "acculturation" into the illusory promises of a "modernization" and "developmentalism" based on the mastery of nature and Indigenous erasure, and have subverted the alleged universality, objectivity, inevitability, and futurity of Western knowledges, lifeways, and global designs.[120] For Mixtec philosopher Ignacio Ortiz Castro (2004), Indigenous peoples have learned existence as resistance. Beyond Eurocentric and anthropocentric linear narratives of "civilization" now in terminal eco-crisis, Indigenous peoples protect their knowledges and lifeways, including cosmovision-based land governance, integrated ecosystem management, and polycultural food and holistic medicine(s). Contemporary Mesoamericans (e.g., Huastecs, Totonacs, Otomis, Chinantecs, Nahuas, Popolucs, Zoques, Mayas) still rely on the local communal cultivation of over 1,330 plant species, including at least 1,140 edible and medicinal species.[121] This impressive diversity results from millennia of Indigenous biocultures, including botany, community-based agroecology, and forest gardening that have nurtured biodiversity, soil fertility, and ecosystem health. For 9,000 years, originary Mesoamericans have nurtured at least 15% of the plant species constituting today's world food system.[122]

Consider Maya land management, including varied forest gardening methods, ranging from domestic forest gardens/sacred groves or *pet kots*[123] to vast *kannan k'aax* landscapes, or "well-cared-for-forests."[124] To untrained Eurocentric eyes, Maya-managed forest landscapes seem "natural" subtropical forests.[125] Nourishing 300 to 600 flowering plants, mostly native species with high nutritional, medicinal, and other ecosystem and use values, *Pet kots* contain "more species of trees, shrubs, bushes, herbs, and grasses than any domestic garden on Earth."[126] Maya forest gardens contain more native species diversity than surrounding forests. Every plant has one or more purposes—for example, medicine, food, material, beauty, or attracting fauna (Campbell 2007). These garden designs nurture insect, bird, and four-legged animal habitats, thereby sustaining and enhancing biodiversity. Another Maya design nurtures complex forest orchards from scratch upon *cenotes* (sinkholes) where water sits at the bottom, trapping soil, and becoming silt-filled, creating hyperhumid microenvironments.[127] Under wet conditions, local "black soils" (*box lum*) rich in organic matter form. Here, Mayas foster over twenty plant species,[128] including sacred cacao trees, shade-grown in mixed polyculture forest gardens including useful native and introduced species.[129] Mayas also create, deepen, and expand other soil depressions to reach rainwater deposits or humidity to raise forest gardens.[130]

Like other Mesoamericans, Mayas also practice bioregionally adapted successional *milpa*-to-forest garden systems. Such human-nurtured ecological succession cycles enable high-performing food landscapes/forests.[131] Beyond simplistic conceptions of Mesoamerican milpas that recognize only beans, maize, and squash with chili and herbs, the milpa actually includes polycultural mixes of hundreds of plants and animal species spread over successional stages and interacting in complementarity—in line with Indigenous cosmovisions. This perennial multi-cropping and multistage cyclical cultivation involve maize (*mays*) and at least ninety other plants.[132] Comparable to Indigenous Turtle Islanders, Mesoamerican (e.g., Maya) forest gardeners introduce species by stages in human-nurtured ecological succession cycles. These start with controlled burns (*yum ik'ob*) by "wind-tenders"[133] to regenerate soil fertility and end with lush regenerated closed-canopy forested landscapes. This cycle, sacralized as embodying the spiritual plenitude of life in its full, thriving, and wholesome diversity and vitality (*Utz'Kaslemal*), fosters (agro)biodiverse landscapes sustainably provisioning communities with food, medicines, and useful materials. An average-sized successional milpa alongside foraged harvests from Indigenous-managed forest gardens meets all nutritional requirements,[134] enabling food sovereignty while fostering forest cover for biodiversity and ecosystem services. While milpas have been characterized as "swidden polyculture agroecosystems" leading to regenerated "secondary forest" which seem natural to outsiders,[135] this description can distract from—and potentially dilute—their underpinning biocultural axiology, Indigenous cosmovision, and social ecology: a regenerative, cyclical, complementary, relational, and reciprocal lifeway.

Indigenous Mesoamericans—including Mayans—are "spiritual caretakers and co-creators" of forest ecosystems.[136] "*Milpa*," from the Nahuatl "*millipan*" means "cultivated place" ("*Kol*" in Yukatek-Maya). The resulting extensive forested landscapes are "*Kannan K'aax*" meaning "well-nurtured-forest garden" or "forest guardian-spirit" conveying sacredness.[137] Such integral ecosystem nurturance proceeds from Mesoamerica's land-based cyclical cosmovisions, which underpin millennia of other complex practices.[138] Much of the ancestral forested lands(capes) of the Yucatan peninsula, Chiapas, and northern Central America result from millennia of Mayan management nurturing vast *kannan k'ax*.[139] Many Indigenous Mesoamericans and Indigenous peoples elsewhere (e.g., Amazonian Kayapó[140] and Bora,[141] and Southeast Asian Dayak[142]) design forest gardens comparable to Maya *pet kots* and *kannan k'ax*. Indigenous resurgence movements continue and reclaim these practices, like the Maya Zapatistas, whose agroecological, decolonizing, and depatriarchalizing project includes Indigenous forest gardening.[143] Throughout Mesoamerica and Abya-Yala, the resurgence of Indigenous identity, territoriality, cosmovision, and communality centers the defense and reconstitution of territories of life, embodying communitarian horizons beyond *glocal* catastrophes.[144]

These examples show that landscapes curated through Indigenous nurturance are neither "wild" nor "pristine," but biocultural landscapes, co-created through mutual human–nonhuman nurturance and congenial design by Indigenous peoples for whom landscapes are spiritual, energy-charged expressions anchored to Mother Earth by land-embedded knowledges.[145] Misrepresenting them as "wilderness" harmfully implies that rich lands and resources emerged "naturally" and lack original caretakers. Such misrepresentations serve those who deny Indigenous and local communities' sovereignty and rights while proclaiming territories and resources "free" for appropriation.[146] Casting Indigenous biocultural landscapes as "wild" is a racist, Eurocentric, and colonial conception erasing Indigenous presence, agency, labor, ingenuity, and sovereignty, and enables colonial dispossession and theft. Reifying myths of empty uncultivated wild lands inhabited by "uncivilized savages" facilitate attempts to legitimize settler/colonial land seizures and the destruction of countless Indigenous nations and ecosystems.[147] The pristine wilderness myth ranges from the doctrine of (Christian) "discovery" and dominion to *terra nullius* and notions of land as underutilized, idle, unimproved, underdeveloped, unproductive, or even unprotected/un-conserved.[148] This buttresses the everyday reproduction of ongoing colonialisms where, building on centuries of land dispossession, states, corporations, and settler societies continue treating stolen Indigenous lands as available for their social reproduction, futurity, settlement, extraction, and "development."[149] From the early colonialisms to present land/water/ocean/green grabs, the erosion of Indigenous ways and land dispossession persists, whether for settler statecraft, extractivism, conservation, purported climate change mitigation, or the advancement of the capitalist "green economy" with its new forms of "renewable violence" so manifest in Indigenous territories.[150] The rupturing of land-based relations is at the root of our crises. Kyle Powys Whyte underlines, "while many . . . are concerned about crossing the ecological tipping point, the relational tipping point got crossed long ago thanks to . . . colonialism, capitalism, and industrialization." Centuries of colonialism made it already "too late for Indigenous climate justice."[151]

NECROPOLITICS OF BIOCULTURAL LOSS: COLONIALISM AND THE GENOCIDE-ETHNOCIDE-ECOCIDE NEXUS

The drivers of the loss of Indigenous lands, knowledges, practices, languages, and cultures converge with those of biodiversity loss. From land-use conversions for extractivist industries, agroindustry, and monocultures to replacing native species and traditional crops with non-native and invasive species, the enclosure and privatization of community-managed lands for state, commercial, and "conservation" purposes drive habitat loss, deforestation, soil erosion, water depletion, climate change, and environmental degradation.[152] This is facilitated by the destructuration of Indigenous customary commons governance in favor of modern/(settler-)colonial state and private property regimes, alongside Eurocentric development, modernization, and industrialization. Such disruptions supplant land-based relational cosmovisions, spiritualties, and knowledges with anthropocentric and patriarchal worldviews that desacralize nonhuman nature and spread Eurocentric onto-epistemologies objectifying and commodifying nature. The notion of "land-as-resource/commodity" facilitates industrial subjugation for extractivism. Large-scale mining, logging, and fossil fuel extraction, agricultural mechanization, and "green revolution" technologies, urbanization, and mega-tourism arrive upon the subordination of land-use and resource governance to state and capitalist pressures. Today, the predominant proximate cause of forest destruction is large-scale commercial farming, accounting for 80% of forest clearance in tropical countries, devastating ecosystems, and Indigenous communities.[153] Often preceding is the undermining of Indigenous biocultures, self-determination, lifeways, health, and socioecological reproduction.[154] Alien(ating) land-uses are imposed through manifest, structural, and socially engineered violence,[155] culturally and physically destroying Indigenous biocultures deemed "obstacles to development."

Dispossession also advances through the coloniality of acculturation/assimilation, demoralizing the intergenerational transmission of Indigenous identity and knowledge—including biocultural heritage—and compelling de-Indigenization. Western, colonial, and modern education—like other apparatuses—has discouraged and punished the reproduction of Indigenous lifeways, languages, education, and (self-)governance. Hegemonic institutions have sought to assimilate Indigenous youth into Christian and modern states, markets, and settler societies, inculcating dogmas of dominion over nature, anthropocentrism, Eurocentric progress, instrumental rationality, resource exploitation, development-as-economic growth, land-as-property, and commodification of life. Coloniality perpetuates modernity's ethnocidal undercurrents wherein Indigenous peoples must abandon their cultures, whiten, and modernize—or die off. Coloniality entails institutionally grooming "accultured" de-Indigenized subjects, including colonially minded intermediaries and so-called postcolonial elites, non-Western only phenotypically but carriers of Eurocentric worldviews and values who mimic and transplant hegemonic languages, lifeways, knowledges, and institutions. This global design has imposed a civilization built on the mastery of nature and linear growth-oriented extraction and accumulation rupturing Mother Earth's metabolic cycles. This clashes with Indigenous cultures whose lifeways/cosmovisions coevolved with land-based cycles and are tailored toward reciprocal regeneration. These regenerative commons-based social ecologies of reciprocal care radically embody living alternatives to hegemonic "progress" and "development."[156] Because land-based communal cultures practice or carry the heritage or remembrance of regenerative lifeways, they stand in the way of the dominating logic's universalist ambitions and are targeted for assimilation or elimination—to be sacrificed at the altar of "progress." Hence the necropolitical intent.

This interlocked subjection of landscapes and Indigenous peoples to domination, imposition, and exploitation brings "ecologically induced genocide,"[157] destroying the intimate inseparability of Indigenous peoples and lands. Consequences include forced migration, demographic, mental, and physical health declines, nutritional/dietary problems, loss of land-based cosmovisions and spiritualities, devaluation of relational knowledges and practices, and a general—often racialized and gendered—aggression against Indigenous peoples, heritages, and knowledges, except when appropriated for state, market, or settler interests. The "death dealing logics"[158] of racism and the ecocide-ethnocide-genocide nexus continue, enabling the "supply chain of violence"[159] lubricating the global political economy. Companies and states, often with international and intergovernmental sponsorship and even "big green" organization's complicity, engage in orchestrated exercises of coercion geared toward the social engineering of compliance with continuing extraction on the penalty of necropolitical violence for any dissent. This often involves structurally violent enclosures, exclusions, encroachments, entrenchments, and manifestly violent terrors.[160] As a continuation and renewal of entrenched cycles of colonization and recolonization, this constitutes a "genocide machine"[161] that normalizes extraction and systematic identity erasure. Industries implicated in violence against Indigenous communities and land dispossession include mining and extractives (including energy), agribusiness, megadams and hydroelectrics, industrial logging, and industrial fishing. "Renewable" energy industries are increasingly complicit, including bio/agrofuel, large-scale wind farm, lithium, rare earth, hydrogen, and solar industries.[162] Indigenous land defenders are disproportionately targeted, communities displaced, and biocultural diversity undermined.

The Environmental Justice Atlas[163] records over 3,000 cases of "ecological distribution conflicts" worldwide over recent decades.[164] Although Indigenous peoples now constitute a small minority, 40% of conflicts since the 1970s pit Indigenous and local communities against extractive, industrial, state, and intergovernmental projects. More than any group, over the last years about 40% of land defenders killed annually have been Indigenous, including 185 deaths in 2017; many more are terrorized and harmed in multiple other ways to violently silence them over racial/(neo)colonial and resource-based conflicts.[165] Conflicts involve fossil fuels, energy, and climate (in)justice, water management, biomass and land, infrastructure and built environment, tourism, mining, nuclear projects, industrial and utilities projects, and green grabbing (including for carbon/biodiversity offsetting and "conservation" enclosures).[166] While the capitalist drive to accumulate, and states' power to enable and enforce it are centrally responsible, the systematic subjection and exposure of specifically identified groups to violence, dispossession, vulnerability, and death necessarily presuppose racial, colonial, and gendered aggression.[167] Certain groups, people(s), and cultures are deemed obstacles to accumulation and enemies of the state because they embody the living heritage of radically different worlds and the proof that another world is possible.

Rodriguez Acha (2017) notes, "[c]apitalism, hetero-patriarchy, colonialism and anthropocentrism operate together, privileging the . . . white heterosexual male to the detriment of other lives and lifeforms." Often Indigenous women lead resistance movements in defense of land, forests, water, ecological and human health, self-determination, and community cohesion against environmental harm, pollution, and extractive industries.[168] Indigenous women are not only disproportionately affected (e.g., due to violence associated with settler extractivist man-camps), but their resistance and defiance are frequently criminalized, stigmatized, and targeted, not least because they are often the carriers of Indigenous biocultural heritage and futures.[169] The prominence and success of Indigenous women's anti-extractivist and environmental/climate activism are rooted in subversive Indigenous land-based spiritualities and communal womanisms in defense of Mother Earth—not Western feminism.[170] Indigenous women's leadership proceeds from Indigenous gender matrices that differ from and have not given into Abrahamic colonial patriarchy; their expertise guides movements for restoring the communal health of all our relations.

Notwithstanding the disproportionate—and frequently lethal—violence faced by Indigenous, Afro-Indigenous, and small-scale peasant and fisher communities, they continue—as they have always done—protecting the lands, waters, oceans, skies, and nonhuman companions deemed sacred relations superior to us. They are our teachers and ancestors on whose wisdom Indigenous communities nurture the world's diversity and polycultures of care constituting the fiber of Indigenous communities since time immemorial—as safeguarded primarily by Indigenous elders and women.

IN LIEU OF CONCLUSION: DECOLONIZATION AND RE-INDIGENIZATION—NOT INCLUSION: (COLONIAL) EQUIVOCATIONS ON "BIOCULTURAL DIVERSITY" AND "INDIGENOUS KNOWLEDGES" IN GLOBAL ENVIRONMENTAL/CLIMATE POLITICS

Indigenous historian Nick Estes (Lakota) raised a key question concerning colonial equivocations in progressive green agendas, like the Green New Deal:

> Why is it easier . . . to imagine the end of fossil fuels but not settler colonialism? To imagine green economies and carbon-free, wind turbine, solar power, and electric bullet train utopias but not the return of Indigenous lands? (2019)

Similar colonial equivocations can be identified in agendas of biocultural diversity and IKs in global environmental/climate politics/policy.

The links between cultural—particularly Indigenous—diversity with biodiversity and terrestrial carbon sinks have brought growing recognition in policy spheres that biodiversity protection, climate change mitigation, and sustainable food systems relate to IKs and rights. Yet, as biocultural diversity and IKs become "policy visible,"[171] their mainstreaming signals colonial equivocations regarding decolonization, Indigenous resurgence, sovereignty, governance, and land rematriation.[172] The biological-cultural link achieved visibility since the 1988 First International Congress of Ethnobiology. Thereafter, several intergovernmental bodies, instruments, and programs started signaling some contributions from IKs to ecosystem vitality, diversity, and resiliency. This is certainly the case for the UN Permanent Forum on Indigenous Issues and the UN Declaration on the Rights of Indigenous Peoples. Yet beyond Indigenous-centered fora, this is notable in the Convention on Biological Diversity (CBD), the UN Framework Convention on Climate Change (UNFCCC), The Millennium Ecosystem Assessment, The International Assessment of Agricultural Science and Technology, and recently in the Intergovernmental Panel on Climate Change, and Intergovernmental Science-Policy Platform on Biodiversity and Ecosystem Services reports.[173]

Such "inclusion" into hegemonic policy-oriented scientific and inter-governmental discourse could be considered an achievement—perhaps tactical. However, while the bioculture-IK-climate link gains attention, the political-ecological demands for Indigenous self-determination, sovereignty, decolonization, and land rematriation in governance, planning, management, and knowledge are often sidestepped, perhaps to avoid unsettling state and market powers. Hence, a logic of "inclusion" where Indigeneity is mixed into dominant and still colonizing frameworks like "market mechanisms" and the state/interstate system often prevails. This threatens or actualizes political neutralization, cooptation, and hegemonic capture.

Inclusionary moves signal colonial equivocation when "Indigenous peoples," "IKs," or "biocultural diversity" are invoked while simultaneously

(1) failing to center decolonizing challenges, critiques, and alternatives to the systems and actors undermining Indigenous peoples and driving global crises—for example, colonialism in all forms, the state/interstate system, capitalism, corporations;
(2) assuming that Indigeneity is available for discursive science and policy representation, even when absent prior commitments to Indigenous anti-colonial, anti-state, anti-extractivist, and anti-capitalist struggles for land restitution, land defense, Indigenous resurgence, sovereign governance, and/or even when absent political/policy and scientific/academic self-representation at the highest levels; and
(3) failing to challenge dominant policy processes, political-economic structures, and scientific frameworks subscribing to or dependent on state/interstate-centric power structures and interests, mostly captured by—or accepting of—capitalist "market mechanisms."[174]

Hegemonic knowledge/science still privileges modern Western onto-epistemologies and has been effectively instrumentalized for state-centered, neoliberal market-driven, and/or technocratic fixes based on reductionist/

Eurocentric scientisms, perpetuating epistemicide and entrenching coloniality and new imperialisms.[175] The predominantly Eurocentric knowledge/science of intergovernmental science-policy discourse is distant from the embodied, grounded, and situated knowledges of Indigenous peoples—including knowledge of resistance and survivance from centuries of ongoing anti-colonial and resurgence struggles. For example, without ignoring the virtues in recent IPCC and IPBES reports, both are still acquiescent to state-centric and market mechanisms. They are still part of UNFCCC and CBD processes geared toward state-centric—and increasingly corporatized—Conferences of Parties (COPs).[176] There, asymmetric power dynamics engender troubling outcomes privileging problematic so-called solutions such as payment for ecosystem services, carbon trading, carbon markets, forest carbon markets, REDD+,[177] carbon/biodiversity offsets, "climate smart agriculture," BECCS,[178] and geo-/bioengineering schemes.[179] Such false "solutions" enclose commons, fail to address environmental crises, harm frontline, local and Indigenous communities and territories, and embody state and corporate interests antagonistic to Indigenous struggles.[180]

"Including" IKs and bioculture while sidelining or deferring the politics of decolonization is like selectively "including" agroecology (e.g., into climate-smart agriculture) while marginalizing food/land sovereignty agendas.[181] IK fragments are selectively recuperated and mixed into hegemonic approaches to conservation and climate change that further commodify and master nature while re-colonizing and displacing Indigenous peoples. This disembeds IKs from struggles to uproot neo-colonialism/settler-colonialism and reclaim Indigenous lands, lifeways, and self-determination/sovereignty. Indigenous decolonizing, anti-extractivist, and often anti-state and anti-capitalist struggles must be foregrounded to disrupt attempts to represent such "moves to IK inclusion" as innocent. This means unsettling the Eurocentric, state-centric, and capitalist power structures upon which hegemonic science and policy frameworks still rely.

Notably, dominant epistemic and policy approaches usually fail to question the sovereign claims of modern nation-states—including settler and so-called "post"colonial states—whose territoriality often results from colonization, dispossession, kidnapping, genocide, and/or forced assimilation of Indigenous peoples, many still struggling to defend and/or recover lands, identities, and self-determination. This bedevils most science-policy frameworks addressing biocultural diversity and Indigenous rights. They sidestep actual decolonization: the dismantling of colonialism in all forms, the return of lands and sovereignty to dispossessed Indigenous peoples and descendants,[182] Indigenous epistemic/cultural/linguistic reclamation,[183] and resurgence of Indigenous lifeways, practices, and governance.[184] Certainly, several promising discourses re-valorizing IKs, practices, land rights, and biocultural approaches to conservation are emerging.[185] But these discourses can—and should—go beyond proposing community-based ecosystem management, to center decolonization and Indigenous resurgence. Otherwise, proposals could be hegemonically recuperated in fragments through state and "market" crisis management strategies. Promising to some extent are Indigenous and Community Conserved Areas (ICCAs) centering Indigenous cosmovision-based "territories of life" (though not without some concerns)[186]; alternatively, consider the Global Forest Coalition's Community Conservation Resilience Initiative.[187] The latter supports bottom-up participatory community conservation in at least sixty-eight communities in twenty-two countries while rejecting state-centric and market-driven approaches and techno-fixes. It proposes redress for lands and resources taken and excising community forests from state reserves to devolve governance to Indigenous communities for ecosystem restoration via IKs.[188] Similarly, the Forest Peoples Programme's customary rights-based approach demands the "return of customary lands taken without prior community consent, including lands obtained by force or deception" and "lands allocated to third parties by the State."[189] This includes "access to justice and creating mechanisms for land restitution" and "restoration" beyond mere compensation to "communities who have suffered land and resource theft."[190]

We must ask how retroactive land restitution would be, given long and ongoing histories of colonization/settler-colonialism, dispossession, assimilation, and ethno/geno/ecocide. While community-based, rights-based, and biocultural approaches do underline the "futility of . . . conventional protected areas" and "market-based approaches,"[191] system change requires transformative oppositional movements for decolonization and Indigenous resurgence. Decolonizing the biocultural approach demands centering the return of colonized and dispossessed lands to Indigenous peoples and the defense and assertion of Indigenous sovereignty—both for lands already held and those to be rematriated to ancestral Indigenous caretakers. For instance, all land which is now claimed by North American settler states was once held by Indigenous Turtle Islanders. Today less than 4% is held by the at least 1,334 surviving recognized and unrecognized Native/First nations of "North America" (barring those in "Mexico"). Materializing the biocultural axiom requires decolonization: going beyond the defense of remaining Indigenous lands to rematriating the lands and sovereignty of "historically" and continuously dispossessed Indigenous peoples. Moreover, new and existing re-Indigenization programs to reclaim Indigenous heritage, practices, lifeways, and communal land-based identities should be scaled out. This means addressing cases where settler colonization and "postcolonial" statecraft reproduce Eurocentric institutions/ideologies, perpetuating dispossession, "internal colonization," and preventing Indigenous peoples from reclaiming lands while distancing those forced to de-Indigenize from reclaiming Indigeneity.

Frameworks that depoliticize, capture, or neutralize the decolonizing, counterhegemonic, subversive, and system-transformative potential of Indigeneity must be rejected in favor of approaches framing IKs, bioculture, and agroecology within the sociopolitical, territorial, and environmental justice preconditions of self-determination, decolonization, land rematriation, cosmovision reclamation, food sovereignty, climate justice, and systemic change. These challenge logics of state and capital, which are based on dominating Indigenous cultures and mastering, desacralizing, objectifying, and commodifying nature—thereby triggering epochal crises. Only thus can the full potential of IKs and Indigenous land governance/management flourish. Since colonialism's onset, Indigenous alternatives have been defended and nurtured, mostly outside and against colonizing and hegemonic institutions. What they embody and propose transcends mainstreamed/institutional frameworks of Indigenous rights and biocultural conservation. While these can be tactically used for Indigenous aims, their mainstreaming embeds colonial equivocations. Either way, Indigenous mobilization for Indigenous aims must proceed apace toward decolonization in spite of, aside from, and beyond institutional obstacles and support. Certainly, Indigenous reclamation is an indispensable pathway to address interlocking and ever-expanding Anthropocene crises. However, resurgence and re-Indigenization are projects worth pursuing in their own—decolonizing—terms.

NOTES

1. C. Bonneuil and J.-B. Fressoz, *The Shock of the Anthropocene* (Verso, 2016).
2. M. Lepori, "There is No Anthropocene: Climate Change, Species-Talk and Political Economy," *Telos* 172 (2015).
3. R.-C. Collard and J. Dempsey, "Accumulation by Difference-Making," *Gender, Place & Culture* 25, no. 9 (2018); G. Di Chiro, "Care Not Growth: Imagining a Subsistence Economy for All," *BJPIR* 21, no. 2 (2019): 303–311; H. Davis and Z. Todd, "On the Importance of a Date—or Decolonizing the Anthropocene," *ACME* 16, no. 4 (2017); L. Pulido, "Racism and the Anthropocene," in *Remains of the Anthropocene* (Chicago: University of Chicago, 2018); J. W. Moore, "The Capitalocene, Part I," *Journal of Peasant Studies* 44, no. 3 (2017); C. G. Gonzalez, "Global Justice in the Anthropocene," in *Environmental Law and Governance for the Anthropocene* (London: Hart, 2017).
4. W. Sachs and T. Santarius, *Fair Future* (London: Zed, 2007).
5. P. E. Perkins, "Climate Justice, Gender, and Intersectionality," in *Routledge Handbook of Climate Justice*, ed. Jafry (Routledge, 2018). For example, 79% of Fortune 500 corporations' board members and 95% of their CEOs are men—mostly white; six of the top ten firms are in fossil fuels.

6. Collard and Dempsey, "Accumulation by Difference-Making," 3–4.

7. Di Chiro, "Care Not Growth"; Davis and Todd, "On the Importance of a Date—or Decolonizing the Anthropocene"; K. Yusoff, *A Billion Black Anthropocenes or None* (University of Minnesota Press, 2019); Pulido, "Racism and the Anthropocene"; Moore, "The Capitalocene, Part I"; J. Davis et al., "Anthropocene, Capitalocene, . . . Plantationocene?," *Geography Compass* 13, no. 5 (2019); M. W. Murphy and C. Schroering, "Refiguring the Plantationocene," *JWSR* 26, no. 2 (2020); B. Walker, "Precarious Time: Queer Anthropocene Futures," *Parrhesia* 30 (2019).

8. Collard and Dempsey, "Accumulation by Difference-Making," 3–4.

9. U. Brand and M. Wissen, "The Imperial Mode of Living as a Major Obstacle to Sustainability Politics," *GAIA* 27, no. 3 (2018).

10. Bonneuil and Fressoz, *The Shock of the Anthropocene*, not-paginated.

11. R. Warlenius et al., "The Concept of 'Ecological Debt': Its Value for Environmental Justice," *Global Environmental Change* 30 (2015): 29.

12. Pulido, "Racism and the Anthropocene."

13. Davis and Todd, "On the Importance of a Date—or Decolonizing the Anthropocene"; Gonzalez, "Global Justice in the Anthropocene."

14. Davis et al., "Anthropocene, Capitalocene, . . . Plantationocene?"; Pulido, "Racism and the Anthropocene."

15. Bonneuil and Fressoz, *The Shock of the Anthropocene*.

16. Ibid; Pulido, "Racism and the Anthropocene."

17. Ibid.; Davis and Todd, "On the Importance of a Date—or Decolonizing the Anthropocene"; M. Knoblauch, "The First Gardeners: Native Americans and New Jersey's Environment at First Contact," *NJS* 5, no. 2 (2019).

18. Moore, "The Capitalocene, Part I."

19. Pulido, "Racism and the Anthropocene," 126.

20. C. Folke et al., "Our Future in the Anthropocene Biosphere," *Ambio* 50 (2021).

21. Davis and Todd, "On the Importance of a Date—or Decolonizing the Anthropocene"; S. Adelman, "Tropical Forests and Climate Change: Critique of Green Governmentality," *International Journal of Law in Context* 11, no. 2 (2015).

22. J. V. Fenelon, "Indigenous Alternatives to the Global Crises of the Modern World-System," in *Overcoming Global Inequalities*, eds. Wallerstein et al. (New York: Routledge, 2015); L. Maffi and E. Woodley, *Biocultural Diversity Conservation* (New York: Earthscan, 2010).

23. Consider the erasing of Indigenous Myaamia (Miami) languages and language reclamation efforts described by Leonard (this volume).

24. V. M. Toledo, "¿Por qué los pueblos indígenas son la memoria de la especie?," *Papeles* 107 (2009).

25. Forest Peoples Programme—FPP, "Central Roles of Indigenous Peoples and Local Communities in Global Commitments on Biodiversity," 2018, https://www.Indigenouspeoples-sdg.org/index.php/english/all-resources/ipmg-position-papers-and-publications/ipmg-submission-interventions/95-the-central-roles-of-Indigenous-peoples-and-local-communities-in-achieving-global-commitments-on-biodiversity/

file; United Nations—UN, *State of the World's Indigenous Peoples* (New York: UN, 2009); World Bank—WB, "Indigenous Peoples," 2019, https://www.worldbank.org/en/topic/Indigenouspeoples#1.

26. First Peoples Worldwide, "Who Are Indigenous Peoples," http://www.firstpeoples.org/who-are-indigenous-peoples.htm.

27. G. Bonfil Batalla, *México Profundo* (Austin: University of Texas, 2004 [1987]).

28. Global Forest Coalition—GFC, "Community Conservation Resilience Initiative Report," 2018, https://globalforestcoalition.org/ccri-global-report/; Global Forest Coalition—GFC, "Policy Recommendations CBD COP 14," 2018, https://globalforestcoalition.org/cbd-cop14-policy-recommendations/.

29. P. K. Austin and J. Sallabank, "Introduction," in *The Cambridge Handbook of Endangered Languages* (Cambridge: Cambridge University Press, 2011).

30. IKs comprise, inter alia, bioculturally diverse Indigenous (i) cosmovisions and spiritualties, (ii) commons-based land tenure/management systems, (iii) food/medicinal systems, (iv) holistic health practices, (v) communal and reciprocal labor regimes, (vi) gender complementarity and/or two-spirit/diverse-worlds identities, (vii) calendar and spatial planning in sync with eco-social cycles, (viii) consensual decision-making systems, (ix) participatory intergenerational place-based education and knowledge transmission, and (x) place-based architecture/design. Finally, Indigenous governance, management, and planning strive to integrate these into the design of convivial self-governing communities that nurture the regeneration of all life, in its full richness, plentiful vitality, and bountiful diversity, for indefinite generations.

31. A. Ford and R. Nigh, *The Maya Forest Garden* (Walnut Creek: Left Coast, 2015).

32. Toledo, "¿Por qué los pueblos indígenas son la memoria de la especie?"

33. Ibid.

34. A. B. M. Vadrot, "Endangered Species, Biodiversity and Politics of Conservation," in *Global Environmental Politics*, ed. Kütting (New York: Routledge, 2018).

35. Adopted by the Convention of Biological Diversity (CBD), LPI measures global biodiversity based on population trends of vertebrate species from marine, freshwater and terrestrial habitats.

36. J. Loh and D. Harmon, *Biocultural Diversity: Threatened Species, Endangered Languages* (Zeist: WWF, 2014); L. Gorenflo, et al., "Co-occurrence of Linguistic and Biological Diversity in Biodiversity Hotspots and High Biodiversity Wilderness Areas," *PNAS* 109, no. 21 (2012); Maffi and Woodley, *Biocultural Diversity Conservation.*

37. R. Phillipson and T. Skutnabb-Kangas, "Linguistic Imperialism and Endangered Languages," in *Handbook of Bilingualism and Multilingualism*, eds. Bhatia and Ritchie (Wiley-Blackwell, 2013).

38. Bonneuil and Fressoz, *The Shock of the Anthropocene*; M. P. Pimbert, *Food Sovereignty, Agroecology and Biocultural Diversity* (Routledge, 2018).

39. K. H. Redford and G. M. Mace, "Conserving and Contesting Biodiversity in the Homogocene," in *Rethinking Environmentalism*, eds. Lele et al. (MIT Press, 2018).

40. Davis and Todd, "On the Importance of a Date—or Decolonizing the Anthropocene."

41. The Red Nation, *The Red Deal: Indigenous Action to Save Our Earth* (Brooklyn: Common Notions, 2021), 20.

42. Leonard (this volume).

43. Indigenous Environmental Network/La Via Campesina/Indigenous Climate Action—IEN/LVC/ICA, et al., "Hoodwinked in the Hothouse," 2021, https://climatefalsesolutions.org/; A. Salleh, "A Materialist Ecofeminist Reading of the Green Economy," in *Routledge Handbook of Transformative Global Studies* (Taylor & Francis, 2020); A. Dunlap and J. Verweijen, "The Evolving Techniques of the Social Engineering of Extraction," *Political Geography* (2021); A. Dunlap, "The Politics of Ecocide, Genocide and Megaprojects," *Journal of Genocide Research* (2020); A. Dunlap, "Wind, Coal, and Copper: Politics of Land Grabbing, Counterinsurgency, and the Social Engineering of Extraction," *Globalizations* 17, no. 4 (2020); B. Sovacool, "Who are the Victims of Low-Carbon Transitions? A Political Ecology of Climate Change Mitigation," *Energy Research & Social Science* 73 (2021); T. Kramarz et al., "Governing the Dark Side of Renewable Energy," *Energy Research & Social Science* 74 (2021); B. Jerez et al., "Lithium Extractivism and Water Injustices in the Salar de Atacama, Chile," *Political Geography* 87 (2021); S. Borras and J. Franco, "Agrarian Climate Justice," 2018, https://www.tni.org/files/publication-downloads/wp_agrarian_climate_justice_final.pdf; Friends of the Earth International—FOEI et al., "Chasing Carbon Unicorns: The Deception of Carbon Markets and 'Net Zero,'" 2021, https://www.foei.org/wp-content/uploads/2021/02/Friends-of-the-earth-international-carbon-unicorns-english.pdf; Friends of the Earth International—FOEI, "Dangers of Bioenergy with Carbon Capture and Storage," 2021, https://www.foei.org/wp-content/uploads/2021/01/Friends-of-the-Earth-International_BECCS_English.pdf; C. Gonzalez, "Climate Change, Race, and Migration," *Journal of Law and Political Economy* (2020); J. L. Fernando, "The Virocene Epoch," *Journal of Political Ecology* 27, no. 1 (2020); L. F. Chaves et al., "Scientists Say Land Use Drives New Pandemics. But What If 'Land' Isn't What They Think It Is?," *PReP Agroecologies*, 2021; H. Walia, *Border and Rule* (Chicago: Haymarket, 2021); Center for Interdisciplinary Environmental Justice—CIEJ, "No Comemos Baterías: Solidarity Science Against False Climate Change Solutions," *Science for the People* 22, no. 1 (Spring 2019).

44. The Red Nation, *The Red Deal*, 11–12.

45. "Syndemic"—combining systemic-*and*-pandemic: systemic convergence of health crises with social/environmental crises.

46. D. McGregor et al., "Indigenous Environmental Justice and Sustainability," *Current Opinion in Environmental Sustainability* 43 (2020); CBD Alliance, "Peoples' Response to the High-Level Summit on Biodiversity," 2020, http://www.cbd-alliance.org/sites/default/files/documents/Peoples%27%20response%20ENG.pdf; M. Stewart-Harawira, "'Re-singing the World.' Indigenous Pedagogies and Global

Crisis," in *Globalization and Minority Cultures*, ed. Crosy (Leiden: Brill, 2015); J. V. Fenelon and J. Alford, "Envisioning Indigenous Models for Social and Ecological Change in the Anthropocene," *JWSR* 26, no. 2 (2020); UFT 2020; Climate Justice Alliance—CJA et al., "Feminist Agenda for a New Green Deal," 2019, http://feministgreennewdeal.com/wp-content/uploads/2019/09/Feminist-GND-Kickstart-note-Final-Draft-9.20.2019.pdf; Walker, "Precarious Time: Queer Anthropocene Futures," 2019; N. Muchhala, "Towards a Decolonial and Feminist Global Green New Deal," 2020, https://www.rosalux.de/en/news/id/43146/towards-a-decolonial-and-feminist-global-green-new-deal; J. Tyberg, "Unlearning: From Degrowth to Decolonization," 2020, https://rosalux.nyc/degrowth-to-decolonization/.

47. Maffi and Woodley, *Biocultural Diversity Conservation*; Caillon et al., "Moving beyond the Human–Nature Dichotomy Through Biocultural Approaches," *Ecology and Society* 22, no. 4 (2017); M. C. Gavin et al., "Defining Biocultural Approaches to Conservation," *Trends in Ecology & Evolution* 30, no. 3 (2015).

48. Forest Peoples Programme—FPP, "Central Roles of Indigenous Peoples and Local Communities in Global Commitments on Biodiversity."

49. J. A. Parrotta and R. L. Trosper, *Traditional Forest-Related Knowledge* (New York: Springer, 2012); Loh and Harmon, *Biocultural Diversity: Threatened Species, Endangered Languages*; Toledo, "¿Por qué los pueblos indígenas son la memoria de la especie?"; Forest Peoples Programme—FPP, "Central Roles of Indigenous Peoples and Local Communities in Global Commitments on Biodiversity."

50. Gavin et al., "Defining Biocultural Approaches to Conservation"; Global Forest Coalition—GFC, "Community Conservation Resilience Initiative Report"; Global Forest Coalition—GFC, "Policy Recommendations CBD COP 14."

51. Gorenflo et al., "Co-occurrence of Linguistic and Biological Diversity in Biodiversity Hotspots and High Biodiversity Wilderness Areas."

52. Toledo, "¿Por qué los pueblos indígenas son la memoria de la especie?"

53. K. Dooley et al., "Missing Pathways to 1.5°C," 2018, www.climatelandambitionrightsalliance.org/report; Forest Peoples Programme—FPP, *Closing the Gap: Rights-Based Solutions for Tackling Deforestation* (Moreton-in-Marsh: FPP, 2018); Friends of the Earth International—FOEI, *Community Forest Management* (Amsterdam: FOEI, 2018).

54. Global Forest Coalition—GFC, "1.5*C from a Community Perspective," *Forest Cover* 57 (2018); M. Gabay and M. Alam, "Community Forestry and Its Mitigation Potential in the Anthropocene," *Forest Policy and Economics* 79 (2016); Adelman, "Tropical Forests and Climate Change: Critique of Green Governmentality"; K. McAfee, "Green Economy and Carbon Markets for Conservation and Development," *International Environmental Agreements* 16 (2016); Friends of the Earth International—FOEI, *Community Forest Management*; D. Cardona-Calle, *Community Forest Management and Agroecology* (Amsterdam: FOEI, 2017).

55. Dooley et al., "Missing Pathways to 1.5°C"; Forest Peoples Programme—FPP, *Closing the Gap: Rights-Based Solutions for Tackling Deforestation*.

56. S.T. Garnett et al., "Spatial Overview of the Global Importance of Indigenous Lands for Conservation," *Nature Sustainability* 1 (2018).

57. Dooley et al., "Missing Pathways to 1.5°C"; Forest Peoples Programme—FPP, "Central Roles of Indigenous Peoples and Local Communities in Global Commitments on Biodiversity"; Global Forest Coalition—GFC, "1.5*C from a Community Perspective."

58. Global Witness, "Enemies of the State? How Governments and Business Silence Land and Environmental Defenders," 2019, https://www.globalwitness.org/en/campaigns/environmental-activists/enemies-state/.

59. IPES-Food & ETC Group, "A Long Food Movement: Transforming Food Systems by 2045," 2021, https://www.etcgroup.org/sites/www.etcgroup.org/files/files/longfoodmovementen.pdf.

60. Dooley et al., "Missing Pathways to 1.5°C"; Gabay and Alam, "Community Forestry and Its Mitigation Potential in the Anthropocene."

61. Global Forest Coalition—GFC, "1.5*C from a Community Perspective"; Forest Peoples Programme—FPP, "Central Roles of Indigenous Peoples and Local Communities in Global Commitments on Biodiversity"; Forest Peoples Programme—FPP, *Closing the Gap: Rights-Based Solutions for Tackling Deforestation*; Dooley et al., "Missing Pathways to 1.5°C"; Global Forest Coalition—GFC, "Community Conservation Resilience Initiative Report"; Global Forest Coalition—GFC, "Policy Recommendations CBD COP 14"; Gabay and Alam, "Community Forestry and its Mitigation Potential in the Anthropocene"; Friends of the Earth International—FOEI, *Community Forest Management*; Cardona-Calle, *Community Forest Management and Agroecology*.

62. Gabay and Alam, "Community Forestry and Its Mitigation Potential in the Anthropocene."

63. Ibid.

64. Ibid; Friends of the Earth International—FOEI, *Community Forest Management*.

65. Indigenous Environmental Network-Climate Justice Alliance—IEN-CJA, "Carbon Pricing: Critical Perspective for Community Resistance," 2017, https://www.ienearth.org/wp-content/uploads/2017/11/Carbon-Pricing-A-Critical-Perspective-for-Community-Resistance-Online-Version.pdf; Minga Indígena, "Climate Chart," 2019, https://350.org/wp-content/uploads/2019/12/CARTACLIMATICA-en.pdf; Dooley et al., "Missing Pathways to 1.5°C."

66. Heinrich-Böll-Stiftung—HBS, "Radical Realism for Climate Justice," 2018, https://www.boell.de/sites/default/files/radical_realism_for_climate_justice_volume_44_all_2.pdf?dimension1=ds_radicalrealism.

67. Maffi and Woodley, *Biocultural Diversity Conservation*; E. Salmón, "Kincentric Ecology: Indigenous Perceptions of Human-Nature Relationships," *Ecological Applications* 10, no. 5 (2000); Gavin et al., "Defining Biocultural Approaches to Conservation"; D. Posey, *Kayapó Ethnoecology and Culture* (London: Routledge, 2002); Knoblauch, "The First Gardeners: Native Americans and New Jersey's Environment at First Contact."

68. Posey, *Kayapó Ethnoecology and Culture*.

69. V. M. Toledo, "Indigenous Peoples and Biodiversity," in *Encyclopedia of Biodiversity*, ed. Levin (Academic Press, 2001); Maffi and Woodley, *Biocultural*

Diversity Conservation; Pimbert, *Food Sovereignty, Agroecology and Biocultural Diversity*; P. Settee, "Indigenous Food Sovereignty in Canada," in *Traditional Ecological Knowledge*, eds. Nelson and Shilling (Cambridge University, 2018); F. K. Lake et al., "Considering Diverse Knowledge Systems in Forest Landscape Restoration," in *Forest Landscape Restoration*, eds. Mansourian and Parrotta (London: Routledge, 2018).

70. Lake et al., "Considering Diverse Knowledge Systems in Forest Landscape Restoration."

71. Ibid.; Forest Peoples Programme—FPP, "Central Roles of Indigenous Peoples and Local Communities in Global Commitments on Biodiversity."

72. V. M. Toledo et al., "Multiple Uses of Tropical Forests by Indigenous Peoples in Mexico," *Conservation Ecology* 7, no. 3 (2003).

73. D. Barton, *Indigenous Agroforestry in Latin America* (Chatham: NRI, 1994); Posey, *Kayapó Ethnoecology and Culture.*

74. Knoblauch, "The First Gardeners: Native Americans and New Jersey's Environment at First Contact."

75. S. Sabbath, "Adaptation, Resilience, and Transformability: Historical Ecology of Traditional Furrow Irrigation System on Slopes of Mount Kilimanjaro" (MA Thesis, Uppsala University, 2015); A. Hemp, "Banana Forests of Kilimanjaro: Biodiversity and Conservation in Chagga Homegardens," in *Forest Diversity and Management*, eds. Hawksworth and Bull (Dordrecht: Springer, 2006).

76. Joshi et al., "Indigenous Systems and Ecological Knowledge among Dayak" (presentation, ICRAF—Southeast Asia, 2004).

77. M. L. Khan et al., "Sacred Groves and Their Significance in Conserving Biodiversity," *International Journal of Ecology and Environmental Sciences* 34, no. 3 (2008).

78. Barton, *Indigenous Agroforestry in Latin America.*

79. Global Forest Coalition—GFC, "Community Conservation Resilience Initiative Report"; Global Forest Coalition—GFC, "Policy Recommendations" CBD COP 14."

80. G. Cajete, *Native Science* (Clear Light Publishers, 2000).

81. Adelman, "Tropical Forests and Climate Change: Critique of Green Governmentality"; M. K. Nelson and D. Shilling, *Traditional Ecological Knowledge* (Cambridge University Press, 2018); Gonzalez, "Global Justice in the Anthropocene"; A. Kothari et al., *Pluriverse: Post-Development Dictionary* (Columbia University Press, 2019).

82. Ganienkeh Council Fire, "Ganienkeh Manifesto," 1974, http://www.ganienkeh.net/manifesto.htm.

83. J. Bacher and D. Beaton, "The Iroquois Speak out for Mother Earth," 2019, http://www.firstnationsdrum.com/2019/04/the-iroquois-speak-out-for-mother-earth/.

84. Ibid.

85. Ibid.

86. Wartman et al., "Temperate Agroforestry: How Forest Garden Systems Combined with People-Based Ethics Can Transform Culture," *Sustainability* 10, no. 7 (2018): 4–7.

87. Ibid.

88. Knoblauch, "The First Gardeners: Native Americans and New Jersey's Environment at First Contact," 181, 183.

89. M. Abrams and G. Nowacki, "Native Americans as Active and Passive Promoters of Mast and Fruit Trees in the Eastern USA," *Holocene* 18, no. 7 (2008); M. Black et al., "Native American Influences on Forest Composition of the Allegheny Plateau," *Canadian Journal of Forest Research* 36, no. 5 (2006); S. Tulowiecki and C. P. S. Larsen, "Native American Impact on Past Forest Composition Inferred from Species Distribution," *Ecological Monographs* 85, no. 4 (2015).

90. Abrams and Nowacki, "Native Americans as Active and Passive Promoters of Mast and Fruit Trees in the Eastern USA," 1126.

91. Ibid.

92. Ibid., 1123.

93. Ibid., 1125.

94. Knoblauch, "The First Gardeners: Native Americans and New Jersey's Environment at First Contact."

95. E. T. Pritchard, "Fire is Our Friend," *Catskill Tri-County Historical Views* (2020); M. Kudish, "The Catskills Burn Index," *Catskill Tri-County Historical Views* 2, no. 1 (Spring 2020).

96. Kudish, "The Catskills Burn Index."

97. Pritchard, "Fire is Our Friend"; Kudish, "The Catskills Burn Index."

98. J. Barker, "Territory as Analytic: Dispossession of Lenapehoking and the Subprime Crisis," *Social Text* 36, no. 2 (2018).

99. Abrams and Nowacki, "Native Americans as Active and Passive Promoters of Mast and Fruit Trees in the Eastern USA," 1126.

100. Knoblauch, "The First Gardeners: Native Americans and New Jersey's Environment at First Contact," 147.

101. Ibid., 146–147, 181–182.

102. K. M. Norgaard, "Colonization, Fire Suppression, and Indigenous Resurgence in the Face of Climate Change," *Resilience*, 2019.

103. D. Rose, *Nourishing Terrains: Australian Aboriginal Views of Landscape and Wilderness* (Australian Heritage Commission, 1996).

104. R. P. Neumann, "Ways of Seeing Africa: Colonial Recasting of African Society and Landscape in Serengeti National Park," *Ecumene* 2, no. 2 (1995).

105. Abrams and Nowacki, "Native Americans as Active and Passive Promoters of Mast and Fruit Trees in the Eastern USA," 1124.

106. Wartman et al., "Temperate Agroforestry: How Forest Garden Systems Combined with People-Based Ethics Can Transform Culture."

107. Northeast Farmers of Color Land Trust, "Who We Are & What We Do," https://nefoclandtrust.org/.

108. Ganienkeh Council Fire, "33 Years Later," http://www.ganienkeh.net/33years/; see also "They Call Us Mohawks" documentary, https://www.youtube.com/watch?v=kQUyrgnqkIg&feature=emb_title.

109. Mohawk Community, "About Us," http://www.mohawkcommunity.com/home/aboutus.html.

110. Three Sisters Sovereignty Project, "About Us," https://www.facebook.com/3SistersSov/about/; Three Sisters Sovereignty Project, "The Three Sisters Sovereign Project," 2020, https://www.tishmancenter.org/blog/the-three-sisters-sovereign-project-part-1; Three Sisters Sovereignty Project, "The Three Sisters Sovereignty Project [Part 2]," 2020, https://www.tishmancenter.org/blog/him92pi7q8ij6ksy8pyv6eq20hui1.

111. Dioganhdih, 2019, https://www.dioganhdih.com/.

112. Cayuga Share Farm, https://cayugasharefarm.org/.

113. Nibezun, "Home," 2021, https://nibezun.org/.

114. Schaghticoke First Nations, "The Schaghticoke Land Reclamation Project," 2021, https://www.schaghticoke.info/conservation-sustainable-developmen.

115. Haudenosaunee Environmental Task Force, "'An Indigenous Strategy for Human Sustainibility,'" http://www.hetf.org/.

116. Ejército Zapatista de Liberación Nacional - EZLN, "300, Part I: A Plantation, a World, a War-Slim Chances," 2018, http://enlacezapatista.ezln.org.mx/2018/08/22/300-part-i-a-plantation-a-world-a-war-slim-chances-subcomandante-insurgente-moises-supgaleano/.

117. Tzul Tzul, "Rebuilding Communal Life: Ixil Women and the Desire for Life in Guatemala," *NACLA Report* 50, no. 4 (2018); G. Tzul Tzul, "Sistemas de Gobierno Comunal Indígena," in *Epistemologías del Sur*, eds. Meneses and Bidaseca (CLACSO, 2018); EDUCA A.C., "Alternativas comunitarias en defensa de los territorios en Oaxaca," 2019, https://www.educaoaxaca.org/wp-content/uploads/2020/01/Final-di%C3%A1gnostico-alternativas.pdf; L. E. Figueroa-Helland, "Indigenous Philosophy and World Politics" (PhD diss., ASU, 2012); B. Maldonado, "Perspectivas de la Comunalidad en Pueblos Indígenas de Oaxaca," *Bajo el Volcán* 15, no. 23 (2016); M.C. Bastida, "Five Hundred Years of Resistance" (MA Thesis, Carleton University, 1997).

118. Ejército Zapatista de Liberación Nacional - EZLN, "300, Part I: A Plantation, a World, a War-Slim Chances."

119. S. Marcos et al., *The Speed of Dreams* (City Lights Publisher, 2007), 81, 80, 45, 90, 120.

120. Bonfil Batalla, *México Profundo*.

121. Toledo et al., "Multiple Uses of Tropical Forests by Indigenous Peoples in Mexico."

122. Toledo et al., "The Biocultural Heritage of Mexico," in *Biocultural Diversity Toolkit*, eds. Maffi and Dilts (Terralingua, 2014).

123. Gómez-Pompa et al., "Sacred Groves of the Maya," *Latin American Antiquity* 1, no. 3 (1990); D.G. Campbell, "Don Berto's Garden," 2007, https://orionmagazine.org/article/don-bertos-garde/.

124. Ford and Nigh, *The Maya Forest Garden*.

125. Campbell, "Don Berto's Garden."

126. For species, see Ibid.

127. Gómez-Pompa et al., "Sacred Groves of the Maya."

128. For species, see Ibid.

129. Ibid.

130. Ibid.

131. Ford and Nigh, *The Maya Forest Garden.*

132. For species, see Ibid.

133. Ibid.

134. Falkowski et al., "More than Just Corn and Calories: Comprehensive Assessment of Yield and Nutritional Content of Traditional Lacandon Maya milpa," *Food Security* 11, no. 2 (2019).

135. Ibid.

136. Ford and Nigh, *The Maya Forest Garden.*

137. Ibid.

138. C. Lenkersdorf, *Conceptos Tojolabales de Filosofía y Altermundo* (Mexico: Plaza y Valdés, 2004); Figueroa-Helland, "Indigenous Philosophy and World Politics"; L. E. Figueroa-Helland and P. Raghu, "Indigeneity vs. 'Civilization'," in *Social Movements and World-system Transformation*, ed. Smith et al. (New York: Routledge, 2016).

139. Ford and Nigh, *The Maya Forest Garden*; Campbell, "Don Berto's Garden."

140. Posey, *Kayapó Ethnoecology and Culture.*

141. Barton, *Indigenous Agroforestry in Latin America.*

142. Joshi et al., "Indigenous Systems and Ecological Knowledge among Dayak."

143. L. Gahman, "Food Sovereignty in Rebellion," *Solutions* 7, no. 4 (2016); Q. Saul, "Resurgent Mexico," 2016, https://dorsetchiapassolidarity.wordpress.com/tag/forest-gardens/; Schools for Chiapas, "El Vivero Muy Otro-Food Forests, Sustainability, and Building a Greenhouse in Chiapas," 2017, https://schoolsforchiapas.org/el-vivero-muy-otro-food-forests-sustainability-and-building-a-greenhouse-in-chiapas/.

144. EDUCA A.C., "Alternativas comunitarias en defensa de los territorios en Oaxaca."

145. Posey, *Kayapó Ethnoecology and Culture.*

146. Ibid.

147. Knoblauch, "The First Gardeners: Native Americans and New Jersey's Environment at First Contact."

148. Pulido, "Racism and the Anthropocene"; S. Vigil, "Green Grabbing-Induced Displacement," in *Handbook on Environmental Displacement and Migration* (New York: Routledge, 2018).

149. Pulido, "Racism and the Anthropocene."

150. Dunlap, "The Politics of Ecocide, Genocide and Megaprojects"; Transnational Institute - TNI, *The Global Land Grab* (Amsterdam: TNI, 2013); Vigil, "Green Grabbing-Induced Displacement"; Global Forest Coalition - GFC, "Community Conservation Resilience Initiative Report"; Global Forest Coalition - GFC, "Policy Recommendations CBD COP 14"; Adelman, "Tropical Forests and Climate Change:

Critique of Green Governmentality"; Friends of the Earth International - FOEI, *Community Forest Management*; A. Alonso-Fradejas, "'Leaving No One Unscathed' in Sustainability Transitions," *Journal of Rural Studies* 81 (2021); D. E. Rocheleau, "Networked, Rooted and Territorial: Green-grabbing and Resistance in Chiapas," *Journal of Peasant Studies* (2015).

151. K. P. Whyte, "Too Late for Indigenous Climate Justice," *WIREs* 11, no. 1 (2020): e603, 1 and 3.

152. Forest Peoples Programme - FPP, *Closing the Gap: Rights-Based Solutions for Tackling Deforestation*; Global Forest Coalition - GFC, "Community Conservation Resilience Initiative Report"; Global Forest Coalition - GFC, "Policy Recommendations CBD COP 14"; Gabay and Alam, "Community Forestry and its Mitigation Potential in the Anthropocene"; Cardona-Calle, *Community Forest Management and Agroecology.*

153. Forest Peoples Programme - FPP, *Closing the Gap: Rights-Based Solutions for Tackling Deforestation.*

154. Forest Peoples Programme - FPP, "Central Roles of Indigenous Peoples and Local Communities in Global Commitments on Biodiversity"; Forest Peoples Programme - FPP, *Closing the Gap: Rights-Based Solutions for Tackling Deforestation*; Global Forest Coalition - GFC, "Community Conservation Resilience Initiative Report"; Global Forest Coalition - GFC, "Policy Recommendations CBD COP 14"; Gabay and Alam, "Community Forestry and its Mitigation Potential in the Anthropocene"; Friends of the Earth International - FOEI, *Community Forest Management*; M. Crook et al., "Ecocide, Genocide, Capitalism and Colonialism," *Theoretical Criminology* 22, no. 3 (2018).

155. Dunlap and Verweijen, "The Evolving Techniques of the Social Engineering of Extraction."

156. N. Estes, "A Red Deal," 2019, https://www.jacobinmag.com/2019/08/red-deal-green-new-deal-ecosocialism-decolonization-Indigenous-resistance-environment; G. Di Chiro, "Care not Growth: Imagining a Subsistence Economy for All," *BJPIR* 21, no. 2 (2019).

157. Crook et al., "Ecocide, Genocide, Capitalism and Colonialism," 299.

158. Davis et al., "Anthropocene, Capitalocene,…Plantationocene?," 3.

159. N. Butt et al., "The Supply Chain of Violence," *Nature Sustainability* 2 (2019).

160. Dunlap and Verweijen, "The Evolving Techniques of the Social Engineering of Extraction"; Dunlap, "The Politics of Ecocide, Genocide and Megaprojects"; Sovacool, "Who are the Victims of Low-carbon Transitions? A Political Ecology of Climate Change Mitigation"; Rocheleau, "Networked, Rooted and Territorial: Green-grabbing and Resistance in Chiapas."

161. Dunlap, "The Politics of Ecocide, Genocide and Megaprojects"; Dunlap, "Wind, Coal, and Copper: Politics of Land Grabbing, Counterinsurgency, and the Social Engineering of Extraction."

162. Global Witness, "Enemies of the State? How Governments and Business Silence Land and Environmental Defenders"; Dunlap, "The Politics of Ecocide, Genocide and Megaprojects"; Dunlap, "Wind, Coal, and Copper: Politics of Land

Grabbing, Counterinsurgency, and the Social Engineering of Extraction"; Indigenous Environmental Network/ La Via Campesina/Indigenous Climate Action - IEN/LVC/ICA, "Hoodwinked in the Hothouse"; Forest Peoples Programme - FPP, "Central Roles of Indigenous Peoples and Local Communities in Global Commitments on Biodiversity"; Forest Peoples Programme - FPP, *Closing the Gap: Rights-Based Solutions for Tackling Deforestation*; Global Forest Coalition - GFC, "Community Conservation Resilience Initiative Report"; Global Forest Coalition - GFC, "Policy Recommendations CBD COP 14"; Gabay and Alam, "Community Forestry and its Mitigation Potential in the Anthropocene."

163. Environmental Justice Atlas, "EJAtlas - Global Atlas of Environmental Justice," https://ejatlas.org/.

164. J. Martinez-Alier et al., "Is there a Global Environmental Justice Movement?," *Journal of Peasant Studies* 43, no. 3 (2016).

165. Butt et al., "The Supply Chain of Violence."

166. Di Chiro, "Care not Growth: Imagining a Subsistence Economy for All"; Adelman, "Tropical Forests and Climate Change: Critique of Green Governmentality"; McAfee, "Green Economy and Carbon Markets for Conservation and Development"; Friends of the Earth International - FOEI, *Community Forest Management.*

167. Pulido, "Racism and the Anthropocene"; A. Perez Aguilera, "Mining and Indigenous Cosmopolitics, The Wirikuta Case," in *Ecological Crisis and Cultural Representation in Latin America*, eds. Anderson and Bora (Lexington Books, 2016).

168. M.A. Rodriguez Acha, "We Have to Wake Up, Humankind! Women's Struggles for Survival, Climate and Environmental Justice," *Development* 60 (2017); International Indigenous Women's Forum - IIWF/FIMI, "Environmental Justice: Perspective of Indigenous Women," 2019, http://www.fimi-iiwf.org/environmental-justice-perspective-of-indigenous-women/?lang=en; M.K. Nelson, "Wrestling with Fire: Indigenous Women's Resistance and Resurgence," *American Indian Culture and Research Journal* 43, no. 3 (2019).

169. Rodriguez Acha, "We Have to Wake Up, Humankind! Women's Struggles for Survival, Climate and Environmental Justice"; Friends of the Earth International - FOEI, *Community Forest Management*; Forest Peoples Programme - FPP, *Closing the Gap: Rights-Based Solutions for Tackling Deforestation*; Global Forest Coalition - GFC, "Community Conservation Resilience Initiative Report"; Global Forest Coalition - GFC, "Policy Recommendations CBD COP 14"; Cardona-Calle, *Community Forest Management and Agroecology*; Colectivo Miradas Críticas del Territorio desde el Feminismo - CMCTF, "(Re)patriarcalización de los territorios. La lucha de las mujeres y los megaproyectos extractivos," *Ecopolitica*, 2018; Sovereign Bodies Institute - SBI, "ZUYA WINYAN WICAYUONIHAN: A Study on Missing & Murdered Indigenous Women and-girls in States Impacted by the Keystone XL Pipeline," 2019, https://2a840442-f49a-45b0-b1a1-7531a7cd3d30.filesusr.com/ugd/6b33f7_27835308ecc84e5aae8ffbdb7f20403c.pdf.

170. Tzul Tzul, "Rebuilding Communal Life: Ixil Women and the Desire for Life in Guatemala"; Tzul Tzul, "Sistemas de Gobierno Comunal Indígena"; Perkins, "Climate Justice, Gender, and Intersectionality"; S. Marcos, "Subversive Spirituality:

Political Contributions of Ancestral Cosmologies Decolonizing Religious Beliefs," in *Dynamics of Religion*, eds. Bochinger and Rupke (Berlin: DeGruyter, 2017).

171. Merçon et al., "From Local Landscapes to International Policy: Contributions of the Biocultural Paradigm to Global Sustainability," *Global Sustainability* 2, no. e7 (2019).

172. K.P. Whyte, "What Do Indigenous Knowledges Do for Indigenous Peoples?," in *Traditional Ecological Knowledge* (Cambridge University, 2018); Di Chiro, "Care not Growth: Imagining a Subsistence Economy for All"; E. Tuck and K.W. Yang, "Decolonization is Not a Metaphor," *Decolonization: Indigeneity, Education & Society* 1, no. 1 (2012); N. Latulippe and N. Klenk, "Making Room and Moving Over: Knowledge Co-production, Indigenous Knowledge Sovereignty and Politics of Global Environmental Change Decision-making," *Current Opinion in Environmental Sustainability* 42 (2019); S. Grey and R. Kuokkanen, "Indigenous Governance of Cultural Heritage," *International Journal of Heritage Studies* 26, no. 10 (2019).

173. Intergovernmental Panel on Climate Change - IPCC, "Global Warming of 1.5°C" (Lecture, Intergovernmental Panel on Climate Change, Geneva, 2018); Intergovernmental Panel on Climate Change - IPCC, "Climate Change and Land" (Lecture, Intergovernmental Panel on Climate Change, Geneva, 2020); Intergovernmental Science-Policy Platform on Biodiversity and Ecosystem Services - IPBES, "Assessment Report on Land Degradation and Restoration" (Lecture, Intergovernmental Science-Policy Platform on Biodiversity and Ecosystem Services, Bonn, 2018); Intergovernmental Science-Policy Platform on Biodiversity and Ecosystem Services - IPBES, "Global Assessment Report on Biodiversity and Ecosystem Services" (Lecture, Intergovernmental Science-Policy Platform on Biodiversity and Ecosystem Services, Bonn, 2019).

174. P. Bond, "Social Movements for Climate Justice during the Decline of Global Governance," in *Rethinking Environmentalism*, ed. Lele et al. (Cambridge: MIT, 2018); G. Martine and J.E. Diniz Alves, "Disarray in Global Governance and Climate Change Chaos," *Revista Brasileira de Estudos de População* 36 (2019); A. Martin et al., "Global Environmental Justice and Biodiversity Conservation," *The Geographical Journal* 179, no. 12 (2013).

175. Adelman, "Tropical Forests and Climate Change: Critique of Green Governmentality"; A. Salleh, "Neoliberalism, Scientism and Earth System Governance," in International Handbook of Political Ecology (Edward Elgar, 2015); German et al., "Environmental Governance: Broadening Ontological Spaces for More Livable Worlds" (Paper, Ostrom Workshop, Indiana University, 2019); Moreno et al., *Carbon Metrics: Global Abstractions and Ecological Epistemicide* (Berlin: Heinrich-Böll-Stiftung, 2016); H.-C. Evans and R. Musvipwa, "The Sustainable Development Goals, Paris Agreement and Addis Agenda," in *Knowledge for Justice* (Cape Town: African Minds, 2017); Carbon Trade Watch - CTW, "Paths Beyond Paris," 2015, http://www.carbontradewatch.org/downloads/publications/PathsBeyondParis-EN.pdf.

176. L. Fuhr et al., "5 Years Later-Happy Birthday, Paris Agreement?," 2020, https://us.boell.org/en/2015/12/15/cop-21-and-paris-agreement-force-awakened; Carbon Trade Watch - CTW, "Paths Beyond Paris"; Friends of the Earth International

- FOEI, "Time to Tackle Biodiversity Loss," 2020, https://www.foei.org/features/draft-cbd-global-biodiversity-framework-assessment; T. Fatheuer, *Disputed Nature: Biodiversity and its Convention* (Berlin: FDCL, 2016).

177. Reducing Emissions from Deforestation and Forest Degradation-plus Managing/Conserving/Enhancing Carbon Stocks.

178. Bioenergy with Carbon Capture and Storage.

179. McAfee, "Green Economy and Carbon Markets for Conservation and Development"; Heinrich-Böll-Stiftung - HBS, "Radical Realism for Climate Justice"; Friends of the Earth International-Green Finance Observatory - FOEI-GFO, "Can Market Based Approaches Tackle Critical Loss of Biodiversity?," 2019, https://www.foei.org/wp-content/uploads/2019/12/Can-market-based-approaches-tackle-critical-loss-of-biodiversity_English.pdf; Friends of the Earth International - FOEI, "Regulated Destruction: Biodiversity Offsetting Enables Environmental Destruction," 2018, https://www.foei.org/wp-content/uploads/2019/11/foe-FoN2-regulated-destruction-EN-WEB.pdf; ETC Group, "The Big Bad Fix: Case against Climate Geoengineering," 2018, https://etcgroup.org/sites/www.etcgroup.org/files/files/etc_bbf_mar2018_us_v1_web.pdf; World Forest Movement - WRM, "What Could be Wrong about Planting Trees?," 2020, https://wrm.org.uy/wp-content/uploads/2020/12/WEB_EN_What-could-be-wrong-about-planting-trees_corrected.pdf.

180. Minga Indígena, "Climate Chart"; Indigenous Environmental Network and Climate Justice Alliance - IEN-CJA, "Carbon Pricing: Critical Perspective for Community Resistance"; T. Osborne et al., "Indigenous Peoples and REDD+: Critical Perspective," *IPCCA* 2014; Vigil, "Green Grabbing-Induced Displacement"; Kothari et al., *Pluriverse: Post-Development Dictionary*.

181. Pimbert, *Food Sovereignty, Agroecology and Biocultural Diversity*; Friends of the Earth International, Transnational Institute, CROCEVIA - FOEI, TNI, CROCEVIA, "'Junk Agroecology': Corporate Capture of Agroecology for Partial Ecological Transitions without Social Justice," 2020, https://www.tni.org/files/publication-downloads/38_foei_junk_agroecology_full_report_eng_lr_0.pdf.

182. Tuck and Yang, "Decolonization is Not a Metaphor."

183. Leonard, this volume.

184. Whyte, "What Do Indigenous Knowledges Do for Indigenous Peoples?"; Latulippe and Klenk, "Making Room and Moving Over: Knowledge Co-production, Indigenous Knowledge Sovereignty and Politics of Global Environmental Change Decision-making."

185. For example, Dooley et al., "Missing Pathways to 1.5°C"; Gavin et al., "Defining Biocultural Approaches to Conservation," 2015; Forest Peoples Programme - FPP, "Central Roles of Indigenous Peoples and Local Communities in Global Commitments on Biodiversity"; Forest Peoples Programme - FPP, *Closing the Gap: Rights-Based Solutions for Tackling Deforestation*; Global Forest Coalition - GFC, "1.5*C from a Community Perspective."

186. ICCAs have not been exempt from concerns; e.g., in Tanzania, ICCAs have been promoted by the same conservation organizations promoting colonial fortress

conservation models that displace and dispossess Maasai and other East African pastoralists (interview).

187. Friends of the Earth International – FOEI, *Community Forest Management*; Global Forest Coalition – GFC, "1.5*C from a Community Perspective"; Global Forest Coalition – GFC, "Community Conservation Resilience Initiative Report"; Global Forest Coalition – GFC, "Policy Recommendations CBD COP 14."

188. Global Forest Coalition - GFC, "Community Conservation Resilience Initiative Report"; Global Forest Coalition - GFC, "Policy Recommendations CBD COP 14."

189. Forest Peoples Programme - FPP, Closing the Gap: Rights-Based Solutions for Tackling Deforestation, 24.

190. Ibid., 46, 49.

191. Global Forest Coalition - GFC, "1.5*C from a Community Perspective."

BIBLIOGRAPHY

Abrams, M. and G. Nowacki. "Native Americans as Active and Passive Promoters of Mast and Fruit Trees in the Eastern USA." *Holocene* 18, no. 7 (2008): 1123–1137.

Adelman, S. "Tropical Forests and Climate Change: Critique of Green Governmentality." *International Journal of Law in Context* 11, no. 2 (2015): 195–212.

Alonso-Fradejas, A. "Leaving No One Unscathed' in Sustainability Transitions." *Journal of Rural Studies* 81 (2021): 127–138.

Austin, P. K. and J. Sallabank. "Introduction." In *The Cambridge Handbook of Endangered Languages*, edited by Austin, P.K. and J. Sallabank, 1–24. Cambridge: Cambridge University Press, 2011.

Bacher, J. and D. Beaton. "The Iroquois Speak out for Mother Earth." 2019. http://www.firstnationsdrum.com/2019/04/the-iroquois-speak-out-for-mother-earth/.

Barker, J. "Territory as Analytic: Dispossession of Lenapehoking and the Subprime Crisis." *Social Text* 36, no. 2 (2018): 19–39.

Barton, D. *Indigenous Agroforestry in Latin America.* Chatham: NRI, 1994.

Bastida, M. C. "Five Hundred Years of Resistance." MA Thesis, Carleton University, 1997.

Black, M., et al. "Native American Influences on Forest Composition of the Allegheny Plateau." *Canadian Journal of Forest Research* 36, no. 5 (2006): 1266–1275.

Bonfil Batalla, G. *México Profundo*. Austin: University of Texas, 2004 [1987].

Bonneuil, C. and J.-B. Fressoz, *The Shock of the Anthropocene*. Verso, 2016.

Borras, S. and J. Franco. "Agrarian Climate Justice." 2018. https://www.tni.org/files/publication-downloads/wp_agrarian_climate_justice_final.pdf.

Brand, U. and M. Wissen. "The Imperial Mode of Living as a Major Obstacle to Sustainability Politics." *GAIA* 27, no. 3 (2018): 287–292.

Butt, N., et al. "The Supply Chain of Violence." *Nature Sustainability* 2 (2019): 742–747.

Caillon, S., et al. "Moving Beyond the Human–Nature Dichotomy Through Biocultural Approaches." *Ecology and Society* 22, no. 4 (2017): 27.

Cajete, G. *Native Science*. Clear Light Publishers, 2000.

Campbell, D. G. "Don Berto's Garden." 2007. https://orionmagazine.org/article/don-bertos-garde/.

Carbon Trade Watch—CTW. "Paths Beyond Paris." 2015. http://www.carbontradewatch.org/downloads/publications/PathsBeyondParis-EN.pdf.

Cardona-Calle, D. *Community Forest Management and Agroecology*. Amsterdam: FOEI, 2017.

Cayuga Share Farm. https://cayugasharefarm.org/.

CBD Alliance. "Peoples' Response to the High-Level Summit on Biodiversity." 2020. http://www.cbd-alliance.org/sites/default/files/documents/Peoples%27%20response%20ENG.pdf.

Center for Interdisciplinary Environmental Justice—CIEJ. "No Comemos Baterías: Solidarity Science against False Climate Change Solutions." *Science for the People* 22, no. 1 (Spring 2019): 33–41.

Chaves, L. F., et al. "Scientists Say Land Use Drives New Pandemics. But What If "Land" Isn't What They Think It Is?" *PReP Agroecologies*, 2021.

Climate Justice Alliance—CJA, et al. "Feminist Agenda for a New Green Deal." 2019. http://feministgreennewdeal.com/wp-content/uploads/2019/09/Feminist-GND-Kickstart-note-Final-Draft-9.20.2019.pdf.

Colectivo Miradas Críticas del Territorio desde el Feminismo—CMCTF. "(Re) patriarcalización de los territorios. La lucha de las mujeres y los megaproyectos extractivos." *Ecopolitica*, 2018.

Collard, R.-C. and J. Dempsey. "Accumulation by Difference-Making." *Gender, Place & Culture* 25, no. 9 (2018): 1349–1364.

Crook, M., et al. "Ecocide, Genocide, Capitalism and Colonialism." *Theoretical Criminology* 22, no. 3 (2018): 298–317.

Davis, H. and Z. Todd. "On the Importance of a Date—or Decolonizing the Anthropocene." *ACME* 16, no. 4 (2017): 761–780.

Davis, J., et al. "Anthropocene, Capitalocene, . . . Plantationocene?" *Geography Compass* 13, no. 5 (2019): e12438.

Di Chiro, G. "Care Not Growth: Imagining a Subsistence Economy for All." *BJPIR* 21, no. 2 (2019): 303–311.

Dioganhdih. 2019. https://www.dioganhdih.com/.

Dooley K., et al. "Missing Pathways to 1.5°C." 2018. www.climatelandambitionrightsalliance.org/report.

Dunlap, A. "The Politics of Ecocide, Genocide and Megaprojects." *Journal of Genocide Research*, 23, no. 2 (2021): 212–235.

———. "Wind, Coal, and Copper: Politics of Land Grabbing, Counterinsurgency, and the Social Engineering of Extraction." *Globalizations* 17, no. 4 (2020): 661–682.

——— and J. Verweijen. "The Evolving Techniques of the Social Engineering of Extraction." *Political Geography*, 2021.

EDUCA A. C. "Alternativas comunitarias en defensa de los territorios en Oaxaca." 2019. https://www.educaoaxaca.org/wp-content/uploads/2020/01/Final-di%C3%A1gnostico-alternativas.pdf.

Ejército Zapatista de Liberación Nacional—EZLN. "300, Part I: A Plantation, a World, a War—Slim Chances." 2018. http://enlacezapatista.ezln.org.mx/2018/08/22/300-part-i-a-plantation-a-world-a-war-slim-chances-subcomandante-insurgente-moises-supgaleano/.

Environmental Justice Atlas. "EJAtlas—Global Atlas of Environmental Justice." https://ejatlas.org/.

Estes, N. "A Red Deal." 2019. https://www.jacobinmag.com/2019/08/red-deal-green-new-deal-ecosocialism-decolonization-Indigenous-resistance-environment.

ETC Group. "The Big Bad Fix: Case against Climate Geoengineering." 2018. https://etcgroup.org/sites/www.etcgroup.org/files/files/etc_bbf_mar2018_us_v1_web.pdf.

Evans, H.-C. and R. Musvipwa. "The Sustainable Development Goals, Paris Agreement and Addis Agenda." In *Knowledge for Justice*, edited by T. Halvorsen, H. Ibsen, H-C. Evans, and S. Penderis, 37–56. Cape Town: African Minds, 2017.

Falkowski, T. B., et al. "More Than Just Corn and Calories: Comprehensive Assessment of Yield and Nutritional Content of Traditional Lacandon Maya Milpa." *Food Security* 11, no. 2 (2019): 389–404.

Fatheuer, T. *Disputed Nature: Biodiversity and its Convention.* Berlin: FDCL, 2016.

Fenelon, J. V. "Indigenous Alternatives to the Global Crises of the Modern World-System." In *Overcoming Global Inequalities*, edited by Wallerstein, et al. New York: Routledge, 2015: 143–167.

——— and J. Alford. "Envisioning Indigenous Models for Social and Ecological Change in the Anthropocene." *JWSR* 26, no. 2 (2020): 372–399.

Fernando J.L. "The Virocene Epoch." *Journal of Political Ecology* 27, no. 1 (2020): 635–684.

Figueroa Helland, L. E. "Indigenous Philosophy and World Politics." PhD diss., ASU, 2012.

Figueroa Helland, L. E. and P. Raghu. "Indigeneity vs. 'Civilization'." In *Social Movements and World-System Transformation*, edited by Smith, et al. New York: Routledge, 2016: 189–211.

First Peoples Worldwide. "Who Are Indigenous Peoples." http://www.firstpeoples.org/who-are-indigenous-peoples.htm.

Folke, C., et al. "Our Future in the Anthropocene Biosphere." *Ambio* 50 (2021): 834–869.

Ford, A. and R. Nigh. *The Maya Forest Garden.* Walnut Creek: Left Coast, 2015.

Forest Peoples Programme—FPP. "Central Roles of Indigenous Peoples and Local Communities in Global Commitments on Biodiversity." 2018. https://www.Indigenouspeoples-sdg.org/index.php/english/all-resources/ipmg-position-papers-and-publications/ipmg-submission-interventions/95-the-central-roles-of-Indigenous-peoples-and-local-communities-in-achieving-global-commitments-on-biodiversity/file.

———. *Closing the Gap: Rights-Based Solutions for Tackling Deforestation.* Moreton-in-Marsh: FPP, 2018.

———. "Local Biodiversity Outlooks 2." 2020. https://www.cbd.int/gbo/gbo5/publication/lbo-2-summary-en.pdf.

Friends of the Earth International—FOEI. *Community Forest Management.* Amsterdam: FOEI, 2018.

———. "Regulated Destruction: Biodiversity Offsetting Enables Environmental Destruction." 2018. https://www.foei.org/wp-content/uploads/2019/11/foe-FoN2-regulated-destruction-EN-WEB.pdf.

———. "Time to Tackle Biodiversity Loss." 2020. https://www.foei.org/features/draft-cbd-global-biodiversity-framework-assessment.

———. "Dangers of Bioenergy with Carbon Capture and Storage." 2021. https://www.foei.org/wp-content/uploads/2021/01/Friends-of-the-Earth-International_BECCS_English.pdf.

——— and Green Finance Observatory (FOEI-GFO). "Can Market Based Approaches Tackle Critical Loss of Biodiversity?" 2019. https://www.foei.org/wp-content/uploads/2019/12/Can-market-based-approaches-tackle-critical-loss-of-biodiversity_English.pdf.

——— and Transnational Institute (TNI), CROCEVIA. "'Junk Agroecology': Corporate Capture of Agroecology for Partial Ecological Transitions Without Social Justice." 2020. https://www.tni.org/files/publication-downloads/38_foei_junk_agroecology_full_report_eng_lr_0.pdf.

———, et al. "Chasing Carbon Unicorns: The Deception of Carbon Markets and 'Net Zero.'" 2021. https://www.foei.org/wp-content/uploads/2021/02/Friends-of-the-earth-international-carbon-unicorns-english.pdf.

Fuhr, L. et al. "5 Years Later—Happy Birthday, Paris Agreement?" 2020. https://us.boell.org/en/2015/12/15/cop-21-and-paris-agreement-force-awakened.

Ganienkeh Council Fire. "Ganienkeh Manifesto." 1974. http://www.ganienkeh.net/manifesto.htm.

Ganienkeh Council Fire. "33 Years Later." http://www.ganienkeh.net/33years/.

Gabay, M. and M. Alam. "Community Forestry and Its Mitigation Potential in the Anthropocene." *Forest Policy and Economics* 79 (2016): 26–35.

Gahman, L. "Food Sovereignty in Rebellion." *Solutions* 7, no. 4 (2016): 77–83.

Garnett, S. T., et al. "Spatial Overview of the Global Importance of Indigenous Lands for Conservation." *Nature Sustainability* 1 (2018): 369–374.

Gavin, M. C., et al. "Defining Biocultural Approaches to Conservation." *Trends in Ecology & Evolution* 30, no. 3 (2015): 140–145.

German, L. et al. "Environmental Governance: Broadening Ontological Spaces for More Livable Worlds." Paper presented at the Ostrom Workshop, Indiana University, 2019.

Global Forest Coalition—GFC. *Forest Cover: 1.5°C from a Community Perspective.* 2018. https://globalforestcoalition.org/wp-content/uploads/2018/12/forestcover-57-web.pdf.

———. "Community Conservation Resilience Initiative Report." 2018. https://globalforestcoalition.org/ccri-global-report/.

———. "Policy Recommendations" CBD COP 14." 2018. https://globalforestcoalition.org/cbd-cop14-policy-recommendations/.

Global Witness. "Enemies of the State? How Governments and Business Silence Land and Environmental Defenders." 2019. https://www.globalwitness.org/en/campaigns/environmental-activists/enemies-state/.

Gómez-Pompa, A., et al. "Sacred Groves of the Maya." *Latin American Antiquity* 1, no. 3 (1990): 247–257.

Gonzalez, C. G. "Global Justice in the Anthropocene." In *Environmental Law and Governance for the Anthropocene*, edited by L. Kotze, 219–240. London: Hart, 2017.

Gonzalez, C. "Climate Change, Race, and Migration." *Journal of Law and Political Economy* 1 no 1 (2020): 109–146.

Gorenflo, L. et al. "Co-occurrence of Linguistic and Biological Diversity in Biodiversity Hotspots and High Biodiversity Wilderness Areas." *PNAS* 109, no. 21 (2012): 8032–803.

Grey, S. and R. Kuokkanen. "Indigenous Governance of Cultural Heritage." *International Journal of Heritage Studies* 26, no. 10 (2019).

Haudenosaunee Environmental Task Force. "An Indigenous Strategy for Human Sustainibility." http://www.hetf.org/.

Heinrich-Böll-Stiftung—HBS. "Radical Realism for Climate Justice." 2018. https://www.boell.de/sites/default/files/radical_realism_for_climate_justice_volume_44_all_2.pdf?dimension1=ds_radicalrealism.

Hemp, A. "The Banana Forests of Kilimanjaro: Biodiversity and Conservation in Chagga Homegardens." *Biodiversity and Conservation* 15 (2006):1193–1217.

Indigenous Environmental Network (IEN) and Climate Justice Alliance (CJA). "Carbon Pricing: Critical Perspective for Community Resistance." 2017. https://www.ienearth.org/wp-content/uploads/2017/11/Carbon-Pricing-A-Critical-Perspective-for-Community-Resistance-Online-Version.pdf.

———, La Via Campesina (LVC) and Indigenous Climate Action (ICA), et al. "Hoodwinked in the Hothouse." 2021. https://climatefalsesolutions.org/.

International Indigenous Women's Forum—IIWF/FIMI. "Environmental Justice: Perspective of Indigenous Women." 2019. http://www.fimi-iiwf.org/environmental-justice-perspective-of-indigenous-women/?lang=en.

Intergovernmental Panel on Climate Change—IPCC. "Climate Change and Land." Lecture given at the Intergovernmental Panel on Climate Change, Geneva, 2020.

———. "Global Warming of 1.5°C." Lecture given at the Intergovernmental Panel on Climate Change, Geneva, 2018.

Intergovernmental Science-Policy Platform on Biodiversity and Ecosystem Services—IPBES. "Global Assessment Report on Biodiversity and Ecosystem Services." Lecture given at the Intergovernmental Science-Policy Platform on Biodiversity and Ecosystem Services, Bonn, 2019.

———. "Assessment Report on Land Degradation and Restoration." Lecture given at the Intergovernmental Science-Policy Platform on Biodiversity and Ecosystem Services, Bonn, 2018.

IPES-Food & ETC Group. "A Long Food Movement: Transforming Food Systems by 2045." 2021. https://www.etcgroup.org/sites/www.etcgroup.org/files/files/longfoodmovementen.pdf.

Jerez, B., et al. "Lithium Extractivism and Water Injustices in the Salar de Atacama, Chile." *Political Geography* 87 (2021).

Joshi, L., et al. "Indigenous Systems and Ecological Knowledge among Dayak." Paper presented at ICRAF—Southeast Asia, 2004.

Khan, M. L. et al. "Sacred Groves and Their Significance in Conserving Biodiversity." *International Journal of Ecology and Environmental Sciences* 34, no. 3 (2008): 277–291.

Knoblauch, M. "The First Gardeners: Native Americans and New Jersey's Environment at First Contact." *NJS* 5, no. 2 (2019): 146–184.

Kudish, M. "The Catskills Burn Index." *Catskill Tri-County Historical Views* 2, no. 1 (Spring 2020).

Kothari, A., et al. *Pluriverse: Post-Development Dictionary*. Columbia University Press, 2019.

Kramarz, T. et al. "Governing the Dark Side of Renewable Energy." *Energy Research & Social Science* 74 (2021): 101902 1–11.

Latulippe, N. and N. Klenk. "Making Room and Moving Over: Knowledge Co-production, Indigenous Knowledge Sovereignty and Politics of Global Environmental Change Decision-Making." *Current Opinion in Environmental Sustainability* 42 (2020): 7–14.

Lake, F. K., et al. "Considering Diverse Knowledge Systems in Forest Landscape Restoration." In *Forest Landscape Restoration*, edited by Mansourian and Parrotta, 37–46. London: Routledge, 2018.

Lenkersdorf, C. *Conceptos Tojolabales de Filosofía y Altermundo*. Mexico: Plaza y Valdés, 2004.

Lepori, M. "There is No Anthropocene: Climate Change, Species-Talk and Political Economy." *Telos* 172 (2015): 103–124.

Loh, J. and D. Harmon. *Biocultural Diversity: Threatened Species, Endangered Languages*. Zeist: WWF, 2014.

Maffi, L. and E. Woodley. *Biocultural Diversity Conservation*. New York: Earthscan, 2010.

Maldonado, B. "Perspectivas de la Comunalidad en Pueblos Indígenas de Oaxaca." *Bajo el Volcán* 15, no. 23 (2016): 151–169.

Marcos, S., et al. *The Speed of Dreams*. City Lights Publishing, 2007.

Marcos, S. "Subversive Spirituality: Political Contributions of Ancestral Cosmologies Decolonizing Religious Beliefs." In *Dynamics of Religion*, edited by Bochinger and Rupke, 109–130. Berlin: DeGruyter, 2017.

Martin, A. et al. "Global Environmental Justice and Biodiversity Conservation." *The Geographical Journal* 179, no. 12 (2013): 122–131.

Martine, G. and J. E. Diniz. "Disarray in Global Governance and Climate Change Chaos." *R.bras.Est.Pop.* 36 (2019): 1–30.

Martinez-Alier, J., et al. "Is there a Global Environmental Justice Movement?" *Journal of Peasant Studies* 43, no. 3 (2016): 731–755.

McAfee, K. "Green Economy and Carbon Markets for Conservation and Development." *International Environmental Agreements* 16 (2016): 333–353.

McGregor, D., et al. "Indigenous Environmental Justice and Sustainability." *Current Opinion in Environmental Sustainability* 43 (2020): 35–40.

Merçon, J., et al. "From Local Landscapes to International Policy: Contributions of the Biocultural Paradigm to Global Sustainability." *Global Sustainability* 2, no. e7 (2019): 1–11.

Minga Indígena. "Climate Chart." 2019. https://350.org/wp-content/uploads/2019/12/CARTACLIMATICA-en.pdf.

Mohawk Community. "About Us." http://www.mohawkcommunity.com/home/aboutus.html.

Moore, J. W. "The Capitalocene, Part I." *Journal of Peasant Studies* 44, no. 3 (2017): 594–630.

Moreno, C., et al. *Carbon Metrics: Global Abstractions and Ecological Epistemicide.* Berlin: Heinrich-Böll-Stiftung, 2016.

Muchhala, N. "Towards a Decolonial and Feminist Global Green New Deal." 2020. https://www.rosalux.de/en/news/id/43146/towards-a-decolonial-and-feminist-global-green-new-deal.

Murphy, M. W. and Schroering, C. "Refiguring the Plantationocene." *JWSR* 26, no. 2 (2020): 400–415.

Nelson, M. K. "Wrestling with Fire: Indigenous Women's Resistance and Resurgence." *American Indian Culture and Research Journal* 43, no. 3 (2019): 69–84.

——— and D. Shilling. *Traditional Ecological Knowledge.* Cambridge University Press, 2018.

Neumann, R. P. "Ways of Seeing Africa: Colonial Recasting of African Society and Landscape in Serengeti National Park." *Ecumene* 2, no. 2 (1995): 149–169.

Newcomb, Steven. "Perspectives: Healing, Restoration, and Rematriation." *News & Notes* (Spring/Summer 1995).

Nibezun. "Home." 2021. https://nibezun.org/.

Norgaard, K. M. "Colonization, Fire Suppression, and Indigenous Resurgence in the Face of Climate Change." *Resilience*, 2019.

Northeast Farmers of Color Land Trust. "Who We Are & What We Do." https://nefoclandtrust.org/.

Ortiz Castro, I. *Acercamiento a la Filosofía y Ética del Mundo Mixteco.* Oaxaca: CONACULTA, 2004.

Osborne, T., et al. "Indigenous Peoples and REDD+: Critical Perspective." *IPCCA*, 2014.

Parrotta, J. A. and R. L. Trosper. *Traditional Forest-Related Knowledge.* New York: Springer, 2012.

Perez Aguilera, A. "Mining and Indigenous Cosmopolitics, The Wirikuta Case." In *Ecological Crisis and Cultural Representation in Latin America*, edited by Anderson and Bora, 179–198. Lexington Books, 2016.

Perkins, P. E. "Climate Justice, Gender, and Intersectionality." In *Routledge Handbook of Climate Justice*, edited by Jafry, 349–358. Routledge, 2018.

Phillipson, R. and T. Skutnabb-Kangas. "Linguistic Imperialism and Endangered Languages." In *Handbook of Bilingualism and Multilingualism*, edited by Bhatia and Ritchie, 495–516. Wiley-Blackwell, 2013.

Pimbert, M. P. *Food Sovereignty, Agroecology and Biocultural Diversity*. Routledge, 2018.

Posey, D. *Kayapó Ethnoecology and Culture*. London: Routledge, 2002.

Pritchard, E. T. "Fire is our Friend." *Catskill Tri-County Historical Views* (2020).

PromoProduxions. "'They Call Us Mohawks' | Ganienkeh—A Way of Life | Turtle Island Trust Documentary 2013." June 23, 2013. YouTube video, 26:00. https://www.youtube.com/watch?v=kQUyrgnqkIg&ab_channel=PromoProduxions.

Pulido, L. "Racism and the Anthropocene." In *Future Remains*, edited by Mitman, G., M. Armiero and R. Emmett Chicago: University of Chicago, 2018: 116–128.

Redford, K. H. and G. M. Mace. "Conserving and Contesting Biodiversity in the Homogocene." In *Rethinking Environmentalism*, edited by Lele, et al., 23–39. MIT Press, 2018.

Rocheleau, D. E. "Networked, Rooted and Territorial: Green-Grabbing and Resistance in Chiapas." *Journal of Peasant Studies* 42 no. 3–4 (2015): 695–723.

Rodriguez Acha, M. A. "We Have to Wake Up, Humankind! Women's Struggles for Survival, Climate and Environmental Justice." *Development* 60 (2017): 32–39.

Rose, D. *Nourishing Terrains: Australian Aboriginal Views of Landscape and Wilderness*. Australian Heritage Commission, 1996.

Sabbath, S. "Adaptation, Resilience, and Transformability: Historical Ecology of Traditional Furrow Irrigation System on Slopes of Mount Kilimanjaro." MA Thesis, Uppsala University, 2015.

Sachs, W. and T. Santarius. *Fair Future*. London: Zed, 2007.

Salleh, A. "A Materialist Ecofeminist Reading of the Green Economy." In *Routledge Handbook of Transformative Global Studies*, edited by S. A. H. Hosseini, J. Goodman, S. C. Motta, B. K. Gills, 247–258. Taylor & Francis, 2020.

———. "Neoliberalism, Scientism and Earth System Governance." In *International Handbook of Political Ecology*, edited by R. L. Bryant, 432-446. Edward Elgar, 2015.

Salmón, E. "Kincentric Ecology: Indigenous Perceptions of Human-Nature Relationships." *Ecological Applications* 10, no. 5 (2000): 1327–1332.

Saul, Q. "Resurgent Mexico." 2016. https://dorsetchiapassolidarity.wordpress.com/tag/forest-gardens/.

Schaghticoke First Nations. "The Schaghticoke Land Reclamation Project." 2021. https://www.schaghticoke.info/conservation-sustainable-development.

Schools for Chiapas. "El Vivero Muy Otro—Food Forests, Sustainability, and Building a Greenhouse in Chiapas." 2017. https://schoolsforchiapas.org/el-vivero-muy-otro-food-forests-sustainability-and-building-a-greenhouse-in-chiapas/.

Settee, P. "Indigenous Food Sovereignty in Canada." In *Traditional Ecological Knowledge*, edited by Nelson and Shilling. Cambridge University, 2018.

Sovacool, B. "Who are the Victims of Low-Carbon Transitions? A Political Ecology of Climate Change Mitigation." *Energy Research & Social Science* 73 (2021): 101916 1–16.

Sovereign Bodies Institute (SBI). "ZUYA WINYAN WICAYUONIHAN: A Study on Missing & Murdered Indigenous Women and-Girls in States Impacted by the

Keystone XL Pipeline." 2019. https://2a840442-f49a-45b0-b1a1-7531a7cd3d30.filesusr.com/ugd/6b33f7_27835308ecc84e5aae8ffbdb7f20403c.pdf.

Stewart-Harawira, M. "'Re-singing the World.' Indigenous Pedagogies and Global Crisis." In *Globalization and Minority Cultures*, edited by Crosy, 160–184. Leiden: Brill, 2015.

The Red Nation. "A Red Deal. Part III." 2020. http://therednation.org/wp-content/uploads/2020/04/Red-Deal_Part-III_Heal-Our-Planet.pdf.

Three Sisters Sovereignty Project. "About Us." https://www.facebook.com/3SistersSov/about.

Three Sisters Sovereignty Project. "The Three Sisters Sovereign Project." 2020. https://www.tishmancenter.org/blog/the-three-sisters-sovereign-project-part-1.

Three Sisters Sovereignty Project. "The Three Sisters Sovereignty Project [Part 2]." 2020, https://www.tishmancenter.org/blog/him92pi7q8ij6ksy8pyv6eq20hui1y.

Toledo, V. M. "¿Por qué los pueblos indígenas son la memoria de la especie?" *Papeles* 107 (2009): 27–38.

———. "Indigenous Peoples and Biodiversity." In *Encyclopedia of Biodiversity*, edited by Levin, 451–463. Academic Press, 2001.

———, et al. "Multiple Uses of Tropical Forests by Indigenous Peoples in Mexico." *Conservation Ecology* 7, no. 3 (2003).

———, et al. "The Biocultural Heritage of Mexico." In *Biocultural Diversity Toolkit*, edited by Maffi and Dilts, 22–31. Terralingua, 2014.

Transnational Institute—TNI. *The Global Land Grab*. Amsterdam: TNI, 2013.

Tuck, E. and K. W. Yang. 2012. "Decolonization is Not a Metaphor." *Decolonization: Indigeneity, Education & Society* 1, no. 1 (2012): 1-40.

Tulowiecki, S. and C.P.S. Larsen. "Native American Impact on Past Forest Composition Inferred from Species Distribution." *Ecological Monographs* 85, no. 4 (2015): 557–581.

Tyberg, J. "Unlearning: From Degrowth to Decolonization." 2020. https://rosalux.nyc/degrowth-to-decolonization/.

Tzul Tzul, G. "Rebuilding Communal Life: IXIL Women and the Desire for Life in Guatemala." *NACLA Report* 50, no. 4 (2018): 404–407.

———. "Sistemas de Gobierno Comunal Indígena." In *Epistemologías del Sur*, edited by Meneses and Bidaseca. CLACSO, 2018.

United Nations—UN. *State of the World's Indigenous Peoples*. New York: UN, 2009.

United Frontline Table (UFT). "A People's Orientation to a Regenerative Economy." 2020. https://www.ienearth.org/wp-content/uploads/2020/06/ProtectRepairInvesBuildTransformDRAFT02.pdf.

Vadrot, A.B.M. "Endangered Species, Biodiversity and Politics of Conservation." In *Global Environmental Politics*, edited by Kütting, 198–226. New York: Routledge, 2018.

Vigil, S. "Green Grabbing-Induced Displacement." In *Handbook on Environmental Displacement and Migration*. New York: Routledge, 2018.

Walia, H. *Border and Rule*. Chicago: Haymarket, 2021.

Walker, B. "Precarious Time: Queer Anthropocene Futures." *Parrhesia* 30 (2019): 137–155.

Warlenius, R., et al. "The Concept of 'Ecological Debt': Its Value for Environmental Justice." *Global Environmental Change* 30 (2015): 21–30.

Wartman, P., et al. "Temperate Agroforestry: How Forest Garden Systems Combined with People-Based Ethics Can Transform Culture." *Sustainability* 10, no. 7 (2018): 2246.

Whyte, K. P. "What Do Indigenous Knowledges Do for Indigenous Peoples?" In *Traditional Ecological Knowledge*. edited by M. K. Nelson and D. Shilling, 57–82. Cambridge University, 2018.

———. "Too Late for Indigenous Climate Justice." *WIREs* 11, no. 1 (2020): e603.

World Bank—WB. "Indigenous Peoples." 2019. https://www.worldbank.org/en/topic/Indigenouspeoples#1.

World Rainforest Movement—WRM. "What Could Be Wrong about Planting Trees?" 2020. https://wrm.org.uy/wp-content/uploads/2020/12/WEB_EN_What-could-be-wrong-about-planting-trees_corrected.pdf.

Yusoff, K. *A Billion Black Anthropocenes or None*. University of Minnesota Press, 2019.

Chapter 2

"The Word for Bringing Bodies Back from Water"

Black Oceanic Ecopoetics and the Re-Imagining of Extinction

Ryan Heryford

In her epic poem, *Zong!*, M. NourbeSe Philip asks, "What is the word for bringing bodies back from water?"[1] Asked now, amidst this melting ice, these rising tides, Philip's question, with its focus on retrieval, recollection, and recovery, may seem at odds with the many contemporary aesthetic depictions of our watery world as containing looming, anticipatory, future-oriented threats.

Covering three-fourths of the globe's surface and containing over 90% of the biosphere, this representational attention to water is not solely rhetorical. From former Maldives president Mohamad Nasheed's sensational cabinet meetings with government ministers held twenty feet beneath the Indian Ocean to Jeff Orlowski's time-lapse videos of crumbling arctic glaciers, aquatic representation finds itself front and center in the imagined future-now of anthropogenic climate change, where gradually deteriorating shorelines evoke what Rob Nixon famously called the "slow violence" of environmental catastrophe, "highlighting disasters that are slow moving and long in the making, disasters that are anonymous . . . attritional and of indifferent interest to the sensation driven technologies of our image saturated world."[2] In her chapter for this collection, Marjolein Oele reads what she calls a "language of the sea" as a counterpoint to terrestrially bound narratives, offering a more expansive and attuned vocabulary for the experience of species-loss, mourning, and rebirth. As Oele and other ecocritics have noted, the lived realities of environmental crisis today urgently challenge us to look beyond the pastoral nostalgia of a "green humanities" and shift our focus toward "blue cultural studies," where the otherworldliness and un-inhabitability of oceanic metaphor might serve as better preparation for the dynamism of these uncertain species futures. Steve Mentz refers to such praxis as “swimmer poetics”:

> To imagine a swimmer poetics for our storm-filled world can generate unsustainable but engaging narratives. Swimmers live in the world and enjoy it, but being in the water means knowing that stability cannot last. As the visible catastrophes of climate change appear, we recognize ourselves in the swimmer more than in the gardener.[3]

Mentz’s thinking marks an important shift from the stasis of Earth-bound linearity to a chaotic and unbalanced fluctuation where dreams of ecological stability are sacrificed for a more nuanced understanding of our relation to environmental change. Yet his celebration of the swimmer reiterates and naturalizes a universalism no different from the earlier environmental writings it seeks to challenge, invoking a species “we” that obscures or renders obsolete the differently allocated forms of violence, privilege, or loss imbued in our aquatic relations.

The question that Philip asks, “what is the word for bringing bodies back from water,” comes not from a shared sense of risk or impermanence, but rather a distinct and particular historical understanding drawn from her epic poem, titled after the 1781 ship *Zong*, which cast over 130 enslaved human beings into the sea for insurance reclamation. Philip’s question reconfigures the poet Amiri Baraka’s earlier reminder that “at the bottom of the Atlantic Ocean there lies a railroad made of human bones.”[4] It envisions the rapid, devastating disruption of lifeworlds, not as a future prospects but as that which continues to be inherited by those surviving the accumulative wakes of colonization, displacement, genocide. The bodies after which Philip asks are specifically black bodies, and the water is specifically ocean water, the Atlantic, where forced passage unmade or extinguished lifeworlds both in and under the belly of the boat, in what Édouard Glissant called “a nonworld from which you cry out . . . a womb, a womb abyss.”[5] Abyss, the word Glissant assigns to ocean depths as well as the hold of the ship, is what he calls a tautology, wherein “the entire ocean, the entire sea gently collapsing in the end into the pleasures of sand, makes one vast beginning, but a beginning whose time is marked by these balls and chains gone green.”[6] The womb-abyss of the ocean and the ship constitutes a social death and fugitive persistence which would ontologically figure as blackness, dialectically producing a modern world born through its relation to anti-blackness. In such a world, as scholars like Saidiya Hartman, Fred Moten, and Jonathan Howard have all noted, “slaves in the hold may be understood to have constituted the ground upon which whiteness could originally stand and purport to be.”[7] This Atlantic abyss, this beginning of lifeworlds predicated on extinction, reminds us that the present epochal predicaments against which we are all faced, unevenly, unequally, have not been caused by all ways of knowing, being, and participating in our shared world.

My work for this collection rests on the preliminary assertion that what is at stake in a vast majority of speculative writings about our imposed existential crisis is not the human at the species level so much as one particular figuration of being, what Sylvia Wynter has referred to as the human as Man.[8] Many anticipatory conversations about extinction not only harken to species collapse but are likewise born from anxieties about the end of whiteness as a prescribed and naturalized form of life, with all its entailed pillars of displacement and genocide, colonization, human enslavement. Mine is certainly not a new intervention, and I am indebted to much recent work in decolonial scholarship and anti-racist environmental justice, including Leonardo Figueroa Helland and Abigail Perez Aguilera's important dissection of the colonial roots underpinning "Anthropocene" etymologies. Likewise, an evergrowing archive of work in eco-cultural anthropology by authors like Anna Tsing, Karen Barad, and Deborah Bird Rose outlines the entangled permeations of Man across our deteriorating ecosystems. Tsing, for instance, envisions Earth as "inherited from Man," emblematized by the plantation, or "ecological simplifications in which living things are transformed into resources—future assets—by removing them from their life worlds," wherein such "machines of replication, ecologies devoted to the production of the same," become "breeding grounds for virulence."[9] For Tsing, Figueroa, and Perez, as well as many others working in decolonial plantationocene studies, land becomes a self-actualized referent to histories of settler-colonial displacement and genocide, refracted into forecasted futures of ecological collapse.

Reading offshore, as a literary and cultural studies scholar working and teaching in a department conventionally called "English," I am interested in the ways by which words, lyric, narrative, and other semantic forms might *tell* about extinction, not as speculative imagining, but as a process of recovery and recuperation of those voices that cannot be translated into the language of Man, a process Philip refers to in *Zong!* as "the inability to tell the story that must be told,"[10] and Glissant likewise calls "a clamor for the right to opacity."[11] The ocean, as a site inextricably linked, yet irreducibly estranged from human habitability, begs for newly correspondent lexicons, syntaxes beyond familiar terrestrial metaphor. Poetry, with its capacity to estrange language, to render words apart from their socially coded automations, and in the words of Audre Lorde, "give name to the nameless so it can be thought,"[12] might help us better address Philip's aquatic question, and in so doing, come to envisage recuperated lifeworlds, possibilities, and pathways before, and also beyond the end of Man.

In the concluding section of *Zong!*, titled "Notanda," a response to the practice and process of her poetics, M. NourbeSe Philip expands her earlier question:

> Months later I do an Internet search for a word or phrase for bringing someone back from underwater that has as precise a meaning as the unearthing contained within the word exhume. I find words like resurrect and subaquatic but not "exaqua." Does this mean that unlike being interred, once you're underwater there is no retrieval—that you can never "exhume" from water? The gravestone or tombstone marks the spot of interment, whether of ashes or the body. What marks the spot of subaquatic death?"[13]

Philip's lament raises questions about the use of spatial coordinates and seemingly stable geographic landmarks to accurately testify to departed lives the historical happenings. As Elizabeth DeLoughrey suggests, writers and artists who seek to enmesh human memory, social time, within patterns of the sea face unique concerns, where "unlike terrestrial space . . . the perpetual circulation of ocean currents means that the sea dissolves phenomenological experience and diffracts the accumulation of narrative."[14] Additionally, as we look toward an oceanic future generated by acidifying seas, such abraded temporalities disavow the geographic methodologies often used to mark epochal shifts, usually determined by stratifications within what Kathryn Yusoff calls "a universalist geologic commons, [which] neatly erases histories of racism that were incubated through the regulatory structure of geologic relations."[15]

For the poet and theorist Christina Sharpe, the unique metaphorical and aesthetic language of the sea already reflects the uneven rhythms and dispersed spatial coordinates of diasporic memory through "the wake," or what she later calls "wake-work," requiring attention to the word's multiple meanings, as a current cast by ships, moving objects, bodies; as ritual performed in collective grieving; as the opening of one's eyes, the coming into consciousness. Drawing from the deep scales of what is called the "residence time" between when a substance enters and leaves the ocean—which, in the case of human blood can last over 260 million years—Sharpe expands upon this aquatic hauntology, writing of

> those Africans . . . who passed through the doors of no return [and] did not survive the holding and the sea . . . [who] are alive in hydrogen, in oxygen, in carbon, in phosphorous, and iron; in sodium and chlorine . . . in the time of the wake, known as residence time.[16]

Bodies as dispersed into oceanic residence time stand less as markers for particular or individual loss than a time-scape that is extinction itself. Atlantic currents, lapping evermore closely to cities and homeplaces, are no prequel to mass death, but rather bring the genocidal violence upon which such worlds were built closer to shore.

The sea, as both a time and materialization of residence, complicates metaphorical readings of the ocean as a state of pure fluidity, uninterrupted repetition, unobstructed circulation. As Tiffany Lethabo King reminds, there is a critical reflex borne from narrative readings across the Black Atlantic to "anticipate that Blackness (people, aesthetics, symbols) will show up as liquidity, fluidity, and flow."[17] Metaphorical placements of black death and genocide within the body of the sea risk obscuring the durational weight of this violence, or even reproducing what Saidiya Hartman calls the spectral pleasures of pain that accompanied chattel slavery.[18] King, instead, signals to the shoal as an indiscernible, liminal, ever-shifting, and transforming terrestrial submersion that defies distinctions between solid and liquid. In her reflections on distributive justice in the wake of Hurricane Katrina, Nancy Tuana likewise envisions a phenomenology that is neither terrestrially static nor fluidly abstract, what she refers to as a viscous porosity:

> Viscosity is neither fluid nor solid, but intermediate between them. Attention to the porosity of interactions helps to undermine the notion that distinctions, as important as they might be in particular contexts, signify a natural or unchanging boundary, a natural kind. At the same time, "viscosity" retains an emphasis on resistance to changing form, thereby a more helpful image than "fluidity," which is too likely to promote a notion of open possibilities and to overlook sites of resistance and opposition or attention to the complex ways in which material agency is often involved in interactions, including, but not limited to, human agency.[19]

The ocean as materializing and manifesting both extra-human agency and human history emerges as viscously porous enmeshment, refusing distinctions between the so-called social and natural. Such ontological muddling likewise complicates metaphorical representations of the sea as either temporally vacant or enveloped in endless cyclicity. Oceanic residency does not obscure history into vacuous fluidity, but rather restructures time apart from linear chronologies of telos and accumulation, evoking instead what Alia Al-Saji refers to as a phenomenology of the durée which "escapes quantification while grounding measure . . . generates intervals through its rhythmic punctuation and hesitation . . . its perceived continuity, or flow, [reliant] on structuring discontinuities and differentiations . . . a kind of multiplicity."[20]

Zong!, which Philip notes was told to her by the spirit of Setaey Adamu Boateng, an ancestral griot whose revelations the poet recovers rather than invents or composes, further extends the spectral porosity of durational residence, as a disembodied voice whose wave-like rhythms take shape within an ever-present, historically weighted, always-encroaching now. Philip writes:

> Our entrance to the past is through memory—either oral or written. And water. In this case salt water. Sea water. And, as the ocean appears to be the same yet is constantly in motion, affected by tidal movements, so too this memory appears stationary yet is shifting always. Repetition drives the event and the memory simultaneously, becoming a haunting, becoming spectral in its nature.[21]

Oceanic memory is thus not a marker, but a movement, a ghostly, unexpected arrival of past contradictions, rearranged and repeated across vast horizons of time. How might poetry as a cultural form attend to such temporal and material patterns of the sea?

In *Zong!* loss is attritional; the rhythmic and spatial assemblage of particular repetitions, including the long gaps of silent, blank space between, creates an ocean of words, dispersed, diffracted, and rearranged across textual currents. Dissolving the word water, Philip evokes the contradictory needs of dying, dehydrated persons, at the same time fearful of the imminent threat of being cast into the sea. In their broken spread across the page, these individuated letters, disinterred from their attached meaning, reorient readers to the spatial relations of *Zong!*, where repetition replicates tidal rhythms familiar yet simultaneously illegible to any fixed method of decipherment. The English word *water*, for instance, dissipates and re-emerges in other variants like "want," "away," "won"; when enmeshed within the current of other letters it becomes "one day," "good," and "go," read in Yoruba, we find the phrases "ó d ábó" (until my/your return) and "ó d ola" (until tomorrow).[22]

The sea, in this case, is not some a priori space whose representation the artist strives to best render in accuracy, but rather an experience of reading. The sea of *Zong!* is a body of words, dispersed, repeated, diffracted across textual currents. Approaching the book as a subaquatic ecosystem might better account for the many contemporary readings of *Zong!*'s unique pedagogical demands, including Evie Shockley's assertion that the story's power relies on its constant flux between arrival and retreat, its refusal to "produce a representation or experience of wholeness out of the voices and fragments,"[23] or Fred Moten's discussion of teaching the text as an aural, communal procession, where "one person can't read this poem; it has to be read symphonically."[24] Philip reflects on her own readings of the text as "an extended ga(s)p, or rather a series of ga(s)ps for air with syllabic sounds attached or overlaid," where the struggle for oxygen that is both signaled by the gasps of drowning persons and the gap*s* between words on the page is "driven pneumatically by the energy of the breath in the spaces that enfold the fragments."[25] She elaborates on this communality of enfolded, fragmented breath in her public readings, writing:

> When I perform Zong!, I allow the words and word clusters to breathe for I 'n I—for the we in us that epigenetically we carry within the memory of our cells. When I invite the audience to read with me, we collectively engage in breathing for the Other—for those who couldn't breathe—then . . .[26]

The community of voices collectively engaged in "breathing for the Other" speaks the text of *Zong!* from markedly different positionalities, accenting the limits and possibilities emergent in these uneven and unequal ecosystems

Zong! #1

w w w w a wa

w a w a t

er wa s

our wa

te r gg g g go

o oo goo d

waa wa wa

w w waa

ter o oh

on o ne w one

w o n d d d

ey d a

dey a ah ay

s one day s

wa wa

Figure 2.1 ***Zong!* #1.** *Source:* p. 3 from *Zong!* © 2008 by M. NourbeSe Philip. Published by Wesleyan University Press and reprinted with permission.

of orality and breath. Critics like Sonya Posmentier have suggested that such experiences invite us to reject a usual logic of considering "natural disaster as 'social disaster' but rather reverse those terms to look at human violence as ecological in a formal sense."[27] I would further argue that such ecological violence is not only imbued in the performative but also permeates the material text of *Zong!*, where the ocean in which bodies drown, dissolve, extinguish, reside apart from recognizably contained corporeal life, is both an ocean of hydrogen, oxygen, sodium, chlorine, as well as a sea made of chalk, kaolin clay, eosin, bromine, and all the elements composing the material body of the published book. As the Atlantic can't be disinterred from the incorporated bodies of centuries-long anti-black genocide, neither can literary entanglement with such bodies of water be separated from the textual, aural, and poetic elements that constitute ontological modes of representing, knowing, and being human in our terrestrial and oceanic worlds. With "in," "water," and "to be" as the three most repeated terms throughout the book, Philip's ecology of subaquatic immersion is inextricably linked to ontology as it is performed and naturalized, estranged or reimagined by words and their uses in varying contexts.

Philip describes this oceanic decomposition of words and semantics as an act of violence not dissimilar from the historical traumas recounted in *Zong!* She writes:

> I murder the text, literally cut it into pieces, castrating verbs, suffocating adjectives, murdering nouns, throwing articles, prepositions, conjunctions overboard, jettisoning adverbs: I separate subject from verb, verb from object—create semantic mayhem, until my hands bloodied, from so much killing and cutting, reach into the stinking, eviscerated innards, and like some seer, sangoma, or prophet who, having sacrificed an animal for signs and portents of new life, or simply life, reads the untold story that tells itself by not telling.[28]

The text that Philip murders in *Zong!* is a legal body, specifically the recorded transcript from the Court of the King's Bench hearings on May 21 and 22, 1783, titled *Gregson v. Gilbert.* The case of the slave-ship *Zong*, originally named *Zorg* (meaning "care" in Dutch), concerned insurance claims made by the Gregson slave-trading syndicate in Liverpool. The ship, which had begun its voyage on August 18, 1781, from Accra to Jamaica, was directed off-course by its captain, Luke Collingwood, who misrecognized the port at Kingston for Saint-Domingue, a navigational error only realized after ten days voyage past the intended harbor. With four days' worth of water rations remaining, Collingwood commanded that upward of 130 enslaved persons be cast overboard. Collingwood, who had no prior experience as a ship captain, had previously worked as a surgeon in West Africa, where, if he deemed an enslaved body unfit for Atlantic passage, would kill that person onshore in what was considered by vesting parties to be a "commercial death" whereby insurance money could be reclaimed. Collingwood recognized that if the enslaved persons aboard the *Zong* died on Jamaican soil, or, from what was in maritime law called "a natural death" from starvation or dehydration, no insurance would be received. However, if the crew could argue that what was deemed in the language of the King's Bench as "cargo" had been destroyed intentionally to preserve a remainder, such mass murder became insurable.

Collingwood died before the shipowners went to court in Liverpool, and all the logs were subsequently destroyed or lost. The only remaining documents from this history of eighteenth-century apocalyptic terror come from the two-day legal proceedings, which despite the protests and advocacy of Olaudah Equiano and Granville Sharpe to make the *Zong* a murder trial, never left the maritime court and remained a case of commercial exchange and insurance.[29] As Ian Baucomb notes, writings on anti-black terror and the ontologies imposed by the legal and commercial discourse of human enslavement continue to draw upon the *Zong* case as a

> black Atlantic allegoresis of the middle passage [that] animates its hauntological interrogation of a classical discourse on justice and exchange by repeatedly posing the question of value as a problem of naming and seeing: of knowing how to name what we see and how to value what we name when we view this event from a distance of two hundred years, or indeed from whatever distance separates the viewing of such an atrocity from this atrocious scene.[30]

Revamping personhood through legal discourses of exchange and commodification, which, much like poetry, relied on the application of specific, intentionally selected words and phrasings, *Gregson v. Gilbert* becomes a pinnacle text in modern constructions of Man, and the *Zong* massacre continues to haunt the oft-unspoken or illegible forms of anti-black violence underwriting late liberalism, a legislative refashioning that Hortense Spillers notes was simultaneously borne within the belly of the boat, where "those African persons in 'Middle Passage' were literally suspended in the 'oceanic, . . . culturally 'unmade,' thrown in the midst of a figurative darkness that 'exposed' their destinies to an unknown course."[31]

Beyond dislodging suspect claims to universalism imposed by Anthropocene discourse, *Gregson v. Gilbert* also requires a rigorous vigilance in addressing the paradigms of what constitutes value, gain, and loss when measuring our shared world, attending to the ways in which the verbal and syntactical underpinnings of such measurements participate in what Frank Wilderson calls the "Settler/Master/Human's grammatical structure."[32] Expanding on

Hartman's theorization of labor, accumulation, racialization, and domination in Black subjectification, King describes the processes and methods by which "Black fungibility" comes to replace labor as the organizing mechanism for value and insurance in the post-Enlightenment world, as "an expression of the gratuitous violence of conquest and slavery whose repertoire has no limits or bounds . . . operat[ing] both materially on the body and produc[ing] Blackness (as idea and symbol) as a discursive space of open possibility."[33] In *Gregson v. Gilbert*, oceanic violence is never held to account, but rather fills the ledgers that determine what constitutes value, what can be insured as loss, or absented entirely from the historical record. Anthropocene extinction, as a discursive process by which self-conscious grammatical structures and vocabularies try to encapsulate previously unimaginable species-loss, draws from an archive of language rooted in the transcripts of the King's Bench, where Man's articulations of value and loss are written over the remainders of anti-black genocide.

Rather than disavowing *Gregson v. Gilbert*, Philip moves within the words of the text itself, conflates, and abstracts them beneath with the watery worlds upon which the hold was suspended and by which she can begin to recover the stories that must and cannot be told. The first twenty-six poems in the book forecast and accent their rearrangement of the language of the King's Bench as a means of highlighting the unaccountable archive, this recalled but never recovered the loss. Each page is bordered by a list of Shona, Twi, and Yoruba names, names that may or may not be those of the massacred, as no names or records of names were kept onboard the ship. As an archive of erasure, *Zong!* qualifies narratives of extinction, insisting on the role of totalizing extinguishment, the complete disintegration of lifeworlds which underwrote the Enlightenment's legislative languages of rights and personhood, the same language of gain and loss that we often use to frame value and stakes in ethical arguments about the Anthropocene today.

The inverse of this language of the law, in the case of the *Zong* massacre, becomes the sea itself, which holds and endlessly repeats a subaltern testimony illegible to the semiotics of commerce and exchange. In *Feeding the Ghosts*, his 1997 novel on the *Zong* massacre, Fred D'Aguiar writes that

> the sea receives a body as if that body has come to rest on a cushion, one that gives way to the body's weight and folds round it like an envelope . . . those bodies have their lives written on salt water. The sea current turns pages of memory.[34]

Like D'Aguiar, Philip's *Zong!* requires one to read the sea to hear the recalled yet irrecoverable voices who evade the limits of language. A

Zong! #22

lives own their facts

of spent lives

murder

market

misfortunes

&

policy

lying dead

under seas

facts own their lives

in circumstance

&

happening

in trial &

declaration

in the absolute

of rule

&

lord

in the absolute

of water

Figure 2.2 ***Zong! #22.*** *Source:* p. 39 from *Zong!* © 2008 by M. NourbeSe Philip. Published by Wesleyan University Press and reprinted with permission.

language of the sea is not simply semiotics of words and their constituents in the world, but rather a way or a practice of reading, where waves and currents disassemble and dissolve familiar letters, signs, sentence structures and phrasings, rearranging them into a poetics apart from recognizable linguistic codes. The voice that emerges is never entirely clear nor even translatable. Much like the ocean surface, it invites our gaze and even reflects it, but obscures that reflection in sporadic and unannounced movement, allowing no transparent look beneath. Such poetic reading practice demands comfort in estrangement, dislocation, and alienation, a receptivity to the alternation of signs and their signifiers, and to what Glissant calls an attendance to opacity, located upon

> the edge of the sea [which] represents the alternation (but one that is illegible) between order and chaos. The established municipalities do their best to manage this constant movement between threatening excess and dreamy fragility. Widespread consent to specific opacities is the most straightforward equivalent of nonbarbarism. We clamor for the right to opacity for everyone.[35]

Glissant clarifies this opacity, not as a romance of abstraction, nor an endorsement of humanist abstruseness over political realization, but rather notes that "the opaque is not the obscure . . . [but] that which cannot be reduced, which is the most perennial guarantee of participation and confluence."[36] Echoing such clamors for irreducibility, Philip offers a poetics that carefully refrains from exoticizing illegibility as if it automatically signaled toward some preexistent truth beyond the purview of language. By working within, and not apart from, the text of *Gregson v. Gilbert*, Philip champions an oceanic poetics capable of rendering opaque the language of the law, property, and personhood so that new forms of relation might emerge.

Philip calls the poems that follow the first twenty-six "translation(s) of the opacity of those early poems—a translation that, like all good translations, has a life of its own."[37] Within these additional poems, we see the language of *Gregson v. Gilbert* reinvented through unique dissections and rearrangements of individual words, as well as a recontextualization of those words within a mixture of different languages, all of which, as Philip notes, were used on board the *Zong*, including Arabic, Dutch, Fon, French, Greek, Hebrew, Italian, Latin, Portuguese, Spanish, Shona, Twi, West African Patois, and Yoruba. Additionally, the broad gaps and spaces of silence permeating the page ask us to allow for words without recognizable meaning, to abandon our desire to find familiar signs as individuated letters expand and contract, near, or distance in proximity. This unguided and uneven mixture of linguistic pairings offers a shared, but not a universal language, attendant to a range of diverse and distinct opacities as they relate to one another in speaking about our fractured and fragmented world.

While reading across multiple languages within and beyond the various combinations and overlapping of words can be alienating, there also emerge within these spaces of abstraction various threads of narrative. One voice, for instance, appears to be that of a sailor complicit in the genocide, writing to a distant character named Ruth, yet simultaneously invoking the storytelling

Figure 2.3 *Source:* p. 181 from *Zong!* © 2008 by M. NourbeSe Philip. Published by Wesleyan University Press and reprinted with permission.

practices of a Yoruba griot, and later a Shona-speaking prisoner that both calls out for water and against the terrors of drowning. An uncomfortable reassessment of Dipesh Chakrabarty's famous assertion that Anthropocene "arises from a shared sense of catastrophe,"[38] Philip's multi-voiced, fragmented narrative combines uneven and unequal personages into a symphonic semblance of terror that refuses to contract or condense toward any one anxiety, pathway, or event. Glissant might call this a "true poetics" whose goal, he writes, is

> to develop everywhere, in defiance of a universalizing and reductive humanism, the theory of specifically opaque structures. In a world of Relation, which takes over from the homogeneity of essence, to accept this opaqueness—that is, the irreducible density of the other—is to truly accomplish, through diversity, a human objective. Humanity is perhaps not the "image of man" but today the evergrowing network of recognized opaque structures.[39]

Poetry and poetics, as "evergrowing network(s) of recognized opaque structures," privilege voice that arises from relation as opposed to essence, a reconfiguration of human, species, affective matter, as born through dispersed connections, rhizomatic entanglements. Sylvia Wynter refers to this as Glissant's movement from an ontogeny of being to that of a *sociogeny* that defines becoming only as coordinated by the relational self's always re-assembling eco-social surrounds.[40] Philip's *Zong!* materializes Glissant's poetic theory as a sociogeny of the Atlantic ocean, a text that is both body and sea, body as the sea, and sea as the textual body massacred, disemboweled, and reassembled across the page. If extinction can speak, if a language exists by which its multidirectional, relationally rippling consequences might be heard, Philip aqua-exhumes such speech within the linguistic sea which is *Zong!* What would it mean, such an oceanic poetics asks, to attend to the lives of others, not because we can definitively locate our own reflective understandings in their vitality or suffering, nor in admiration and humanist romanticizing of their indecipherability, but only as we hear the place of their relation in constituting and sustaining a world we share, unevenly, unequally?

In a poem published fifty-five years prior to *Zong!*, Glissant meditates on this question as well as the question with which I opened this essay:

> The ancestor speaks, it is the ocean, it is a race that washed the continents with its veil of suffering; it says this race which is song, dew of song and the muffled perfume and the blue of the song, and its mouth is the song of all the mouths of foam: ocean! you permit, you are accomplice, maker of the stars; how is it you do not open your wings into a voracious lung? And see! there remains only the sum of the song and the eternity of voice and childhood already of those who will inherit it. Because as far as suffering is concerned it belongs to us all; everyone has its vigorous sand between their teeth. The ocean is patience, its wisdom is the tare of time.[41]

One among many of his collected poems dedicated to "every country . . . diverted from its course, [who] suffers the failing of waters,"[42] Glissant's "Ocean" appears more prophetic than ancestral in its invocation. It permits and acts, complicit in the wash and rinse of continents, and if it is patient, it is also vigorous and connective in its suffering. The ocean's wisdom, Glissant concludes with something of a riddle, is "the tare of time." In this English translation of Glissant's poem, the word "tare" has been rendered by Jeff Humphries and Melissa Manolas from the French *l'ivraie*, and could be read as either a form of madness or insanity, or a flaw, a deficit, a signaling toward processes of degeneration. The translated English word might also be traced to its Middle French origins as *tare*, which is often used in the language of commerce to signify tare weight or the weight of the shipping container usually subtracted from the gross value of its presumably more precious interiors. Wisdom, as embodied within Glissant's ocean, is both that which carries and that which destroys. It is the container wherein we ship the weight of the world, as well as the injurious weed sown among the grain that will ultimately destroy the whole of the harvest. To reconcile this interwoven wisdom as bearer and destroyer, connecter and eraser, requires us to embrace a black ecopoesis of the ocean as a practice of sociogenic relation, expanding and contracting along the contours of our own pending species extinction.

Life sustaining and evermore displacing, Glissant's ocean patiently celebrates a degeneration in the globalizing networks attributed to its increasing acidification and rising shorelines. Its endlessly proliferating diffusion offers counter-resolution to Maurice Merleau-Ponty's lament that our other, worldly self can never be fully realized, insofar as "the person who perceives is not spread out before himself as a consciousness must be, [but] has historical density."[43] Moreover, the ocean's fluid tactility, its "porous viscosity," is imbued with a painful temporal collapsing signifying the long durée of white supremacist violence. As Al-Saji writes:

> We are reminded that the very duration of colonialism and white supremacy makes a difference: that they intensify through time, even while being rephrased. Its "retrograde movement," or feedback loop, institutes a history that naturalizes and justifies colonial conquest by scapegoating the bodies and cultures of those who came to be colonized. But this is also a duration that needs to be shored up and maintained by active forgetting and disregard in the present and by reiterations and reinventions of colonial formations through other means. For the colonized and racialized—or the "formerly" colonized—to live under the weight of what I am calling colonial durée (colonial duration) is to experience a "painful sense of time."[44]

Glissant's ocean, which like the undecipherable story that must be told in Philip's *Zong!*, is only the remainder which is "the sum of the song." Yet such remainders carry with them the temporal eviscerations of linear colonial histories, refracting into anticipatory visions of extinction the weight of this "painful sense of time."

To immerse oneself within the ocean's "mouth of foam" is always and necessarily an examination of the weight of the various framing devices through which we come to transport, interpret, and legitimize the knowledge and representation of our planetary connections, an invitation to incorporate lost possibilities into our recuperations of histories which cannot but must be told. As feminist philosopher Veena Das asks in her meditations on historical justice, "are there other paths on which self-creation may take place, through occupying the same place of devastation yet again, by embracing the signs of injury and turning them into ways of becoming subjects?"[45] In *Lose Your Mother*, Saidiya Hartman expands on Das' question within oceanic markers of mourning, inviting, and lamenting the contradictions that arise as lost life-worlds signal toward future possibility. Hartman writes:

> When does one decide to stop looking to the past and instead conceive of a new order? When is it time to dream of another country or to embrace other strangers as allies or to make an opening, an overture, where there is none? When is it clear that the old life is over, a new one has begun, and there is no looking back? From the holding cell was it possible to see beyond the end of the world and to imagine living and breathing again?[46]

The readings I have tried to outline throughout this short chapter conflate Hartman's opening question within the body of the sea, where oceanic currents dissolve and abandon linear models of time, the future spreads across a tideline lapping evermore closely up the shore, carrying the recalled but not recovered voices of centuries-old lifeworlds drowned in the name of a language which cannot hear them. To speak of extinction in this language, to name the collapse of species in the Anthropocene by the terms of value and loss, accumulation and exchange, natural and commercial death, is, as Glissant and Philip remind, always already a failed project. The ocean as a poetics, a semiotics of recall and response, offers another way of being in the world, where the struggle for climate justice is not apart from but inherited by and integrally entangled within the spectral hauntings of past genocides left unresolved. The sea as an extinction event already occurred, looms relentlessly, offering an ever-present now that speaks before and beyond the limits set by the language of Man, demanding what Frantz Fanon once famously called the "re-introduction of invention into existence."[47] In his poem, Glissant asks the ocean "how is it you do not open your wings into a voracious lung?" To hear this lung would require a new type of human, capable of listening to oceanic illegibility. Not simply romanticizing illegibility as inherently transgressive, but rather developing a more active poetics of relation in attendance to the challenges these ruptures entail. Because we will want to hear it, as the wings do open, and the voracious lung projects.

NOTES

1. M. NourbeSe Philip, *Zong!* (Middleton: Wesleyan Poetry Series Press, 2011), 201.
2. Rob Nixon, *Slow Violence and the Environmentalism of the Poor* (Cambridge, MA: Harvard University Press, 2011), 3.
3. Steve Mentz, "After Sustainability," *PMLA* Vol. 127, No. 3 (2012), 590.
4. Amiri Baraka, *Transbluesency: The Selected Poems of Amiri Baraka/Leroy Jones (1961–1995)* (New York: Marisol Publications, 1995), 82.
5. Édouard Glissant, *The Poetics of Relation.* Trans. Betsy Wing (Ann Arbor: University of Michigan Press, 1997), 6.
6. Ibid., 6.
7. Jonathan Howard, "The Soles of Black Folk: Blackness and the Lived Experience of Relation," *ESU Review* (2016): 17.
8. Sylvia Wynter, "Beyond Miranda's Meanings: Un/Silencing the 'Demonic Ground' of Caliban's 'Woman,'" in *Out of the Kumbla: Caribbean Women and Literature* (Africa World Press, 1990), 370.
9. Anna Tsing, "Earth Stalked By Man," *The Cambridge Journal of Anthropology* Vol. 34, No. 1 (Spring 2016), 12–14.
10. Philip, *Zong!*, 2.
11. Glissant, *The Poetics of Relation,* 212–13.
12. Audre Lorde, "Poetry Is not a Luxury," in *Sister Outsider: Essays and Speechees* (Berkeley, CA: Crossing Press, 1985), 36.
13. Philip, *Zong!*, 201.
14. Elizabeth DeLoughrey, *Routes and Roots: Navigating Caribbean and Pacific Island Literature* (Honolulu: University of Hawai'i Press, 2007), 55.
15. Kathryn Yusoff, *A Billion Black Anthropocenes or None* (Minneapolis: University of Minnesota Press, 2019), 14.
16. Christina Sharpe, *In the Wake: On Blackness and Being* (Durham: Duke University Press, 2016), 19.
17. Tiffany Lethabo King, *The Black Shoals: Offshore Formations of Black and Native Studies* (Durham: Duke University Press, 2019), 8.
18. See Saidiya V. Hartman, *Scenes of Subjection: Terror, Slavery and Self-Making in Nineteenth-Century America* (Oxford: Oxford University Press, 1997), 17–49.
19. Nancy Tuana, "Viscous Porosity: Witnessing Katrina," in *Material Feminisms*, ed. Stacy Alaimo and Susan Hekman (Bloomington: Indiana University Press, 2008), 193–4.

20. Alia Al-Saji, *"Durée" 50 Concepts for a Critical Phenomenology*, ed. Gail Weiss, Ann V. Murphy, Gayle Salamon (Chicago: Northwestern University Press, 2019), 100.

21. Philip, *Zong!*, 201.

22. Philip provides a "Glossary" for *Zong!* which includes what she says are "Words and Phrases Overheard On Board the *Zong*" as well as an appendix, titled "Manifest" which includes repeated words and phrasings as grouped into the follow categories: "African Groups and Languages; Animals; Body Parts; Crew; Food and Drink; Nature; Women Who Wait" (pp. 183–6).

23. Evie Shockley, "Going Overboard: African American Poetic Innovation and the Middle Passage," *Contemporary Literature* Vol. 52, No. 4 (2011), 795.

24. As cited in Sonya Posmentier, *Cultivation and Catastrophe: The Lyric Ecology of Modern Black Literature* (Baltimore: Johns Hopkins University Press, 2017), 23.

25. M. NourbeSe Philip, *"The Ga(s)p." Poetics and Precarity (The University at Buffalo Robert Creeley Lectures in Poetry and Poetics)*. Ed. Myung Mi Kim and Cristanne Miller (Buffalo: State University of New York Press, 2018), 39.

26. Ibid., 39.

27. Posmentier, *Cultivation and Catastrophe*, 33.

28. Philip, *Zong!,* 193–4.

29. For more on the history of the *Zong* massacre, see James Walvin's *Zong: A Massacre, the Law and the End of Slavery* (2011) and Marcus Rediker's *The Slave Ship: A Human History* (2008).

30. Ian Baucom, *Specters of the Atlantic: Finance Capital, Slavery, and the Philosophy of History* (Durham: Duke University Press, 2005), 63.

31. Hortense Spillers, *"Mama's Baby, Papa's Maybe' Black, White, and in Color: Essays on American Literature and Culture* (Chicago: University of Chicago Press, 1987), 72.

32. Frank B. Wilderson III, *Red, White, and Black: Cinema and the Structure of U.S. Antagonisms* (Durham: Duke University Press, 2010), 207.

33. King, *The Black Shoals*, 23.

34. Fred D'Aguiar, "From *Feeding the Ghosts*" Paper Airplane (N. 30, 1998), 4.

35. Glissant, *The Poetics of Relation*, 212–13.

36. Ibid., 190–1.

37. Philip, *Zong!*, 206.

38. Dipesh Chakrabarty, "The Climate of History: Four Theses," *Critical Inquiry* Vol. 35, No. 2 (Winter 2009), 7.

39. Glissant, *The Poetics of Relation*, 133.

40. "It is this imperative of a shift from ontogeny to sociogeny, from *l'être* to *l'étant*, and the new frontiers of being and knowing that such a shift opens, that is to be, I believe, the gift of the New World to the Old, the gift specifically of that Other America, that Antilia of both Toscanelli's and Columbus's imaginary geography of some five centuries ago; today the Antilia of Glissant's dream for a fully realized archipelago, for the *avenir* of its small countries, for its collective free, as his oeuvre incites them to be, in their acts, in their desire." Wynter, "Beyond the World of Man," 646.

41. Glissant, *The Collected Poems of Édouard Glissant*. Ed. Jeff Humphries and Jane Humphries. Trans. Melissa Manolas (Minneapolis: University of Minnesota Press, 2005), 50.

42. Glissant, "Eulogy for the Different and Difference," Opening Speech for the 6th International Literature Festival, Berlin, September 5th 2006, Haus der Berliner Festspiele (http://www.literaturfestival.com/intern/reden/glissant-engl).

43. Maurice Merleau-Ponty, "What is Phenomenology?" *Cross Currents*. trans. John F. Bannon Vol. 6, No. 1 (Winter, 1956), 60.

44. Al-Saji, *"Durée" 50 Concepts for a Critical Phenomeno*, 100.

45. Veena Das, *Life and Words: Violence and the Descent Into the Ordinary* (University of California Press, 2006), 215.

46. Saidiya Hartman, *Lose Your Mother: A Journey Along the Atlantic Slave Route* (New York: Farrar, Strauss and Giroux, 2008), 98.

47. Frantz Fanon, *The Wretched of the Earth* (New York: Grove Press, 2011), 38.

BIBLIOGRAPHY

Al-Saji, Alia. *"Durée" 50 Concepts for a Critical Phenomenology*. Ed. Gail Weiss, Ann V. Murphy, Gayle Salamon. Chicago: Northwestern University Press, 2019.

Baraka, Amiri. *Transbluesency: The Selected Poems of Amiri Baraka/Leroy Jones (1961–1995)*. New York: Marisol Publications, 1995.

Baucom, Ian. *Specters of the Atlantic: Finance Capital, Slavery, and the Philosophy of History*. Durham: Duke University Press, 2005.

———. "The Human Shore: Postcolonial Studies in the Age of Natural Science." *History of the Present* 2, no. 1 (May 15, 2012), pp. 1–23.

Chakrabarty, Dipesh. "The Climate of History: Four Theses." *Critical Inquiry* 35, no. 2 (Winter 2009), pp. 197–222.

DeLoughrey, Elizabeth. *Routes and Roots: Navigating Caribbean and Pacific Island Literature*. Honolulu: University of Hawai'i Press, 2007.

———. "Submarine Futures of the Anthropocene." *Comparative Literature* 69, no. 1 (2017), pp. 32–44.

Fanon, Frantz. *The Wretched of the Earth*. New York: Grove Press, 2011

Glissant, Édouard. *Caribbean Discourse: Selected Essays*. Trans. Michael Dash. Charlottesville: University Press of Virginia, 1999.

———. "Eulogy for the Different and Difference." Opening Speech for the 6th International Literature Festival, Berlin, September 5th 2006, Haus der Berliner Festspiele. (http://www.literaturfestival.com/intern/reden/glissant-engl)

———. *The Collected Poems of Édouard Glissant*. Ed. Jeff Humphries and Jane Humphries. Trans. Melissa Manolas. Minneapolis: University of Minnesota Press, 2005.

———. *The Poetics of Relation*. Trans. Betsy Wing. Ann Arbor: University of Michigan Press, 1997.

Hartman, Saidiya. *Lose Your Mother: A Journey Along the Atlantic Slave Route*. New York: Farrar, Strauss and Giroux, 2008.

———. *Scenes of Subjection: Terror, Slavery and Self-Making in Nineteenth-Century America*. Oxford: Oxford University Press, 1997.

Humphries, Jeff. "Introduction." In *The Collected Poems of Édouard Glissant*. Ed. Jeff Humphries and Jane Humphries. Trans. Melissa Manolas, 3–22. Minneapolis: University of Minnesota Press, 2005.

King, Tiffany Lethabo. *The Black Shoals: Offshore Formations of Black and Native Studies*. Durham: Duke University Press, 2019.

Lorde, Audre. "Poetry Is Not a Luxury." In *Sister Outsider: Essays and Speechees*. Berkeley, CA: Crossing Press, 1985, pp. 36–40.

Mentz, Steve. "After Sustainability." *PMLA* 127 3 (2012), pp. 586–92.

Merleau-Ponty, Maurice. 'What is Phenomenology?' *Cross Currents*. trans. John F. Bannon 6, no. 1 (Winter, 1956), p. 60.

Moten, Fred. "Blackness and Nothingness (Mysticism in the Flesh)." *South Atlantic Quarterly* 112, no. 4 (2013), pp. 737–80.

———. "Blackness and Poetry." *Evening Will Come: A Monthly Journal of Poetics* 55 (2015) https://www.thevolta.org/ewc55-fmoten-p1.html.

———. *Social Text*. Thirtieth Anniversary Panel Discussion. Social Text Collective, November 13, 2009, New York.

Nixon, Rob. *Slow Violence and the Environmentalism of the Poor*. Cambridge, MA: Harvard University Press, 2011.

Philip, M. NourbeSe. "The Ga(s)p." In *Poetics and Precarity (The University at Buffalo Robert Creeley Lectures in Poetry and Poetics)*. Ed. Myung Mi Kim and Cristanne Miller. Buffalo: State University of New York Press, 2018.

———. *Zong!* Middleton: Wesleyan Poetry Series Press, 2011.

Posmentier, Sonya. *Cultivation and Catastrophe: The Lyric Ecology of Modern Black Literature*. Baltimore: Johns Hopkins University Press, 2017.

Sharpe, Christina. *In the Wake: On Blackness and Being*. Durham: Duke University Press, 2016.

Shockley, Evie. "Going Overboard: African American Poetic Innovation and the Middle Passage." *Contemporary Literature* 52, no. 4 (2011), pp. 791–817.

Spillers, Hortense J. *"Mama's Baby, Papa's Maybe' Black, White, and in Color: Essays on American Literature and Culture*. Chicago: University of Chicago Press, 1987.

Tsing, Anna. "Earth Stalked By Man." *The Cambridge Journal of Anthropology* 34, no. 1 (Spring 2016), pp. 2–16.

Tuana, Nancy. "Viscous Porosity: Witnessing Katrina." In *Material Feminisms*. Ed. Stacy Alaimo and Susan Hekman. Bloomington: Indiana University Press, 2008.

Weheliye, Alexander G. *Habeas Viscus: Racializing Assemblages, Biopolitics, and Black Feminist Theories of the Human*. Duke University Press, 2014

Wilderson, Frank B., III. *Red, White, and Black: Cinema and the Structure of U.S. Antagonisms*. Durham: Duke University Press, 2010.

Wynter, Sylvia. "Beyond the World of Man: Glissant and the New Discourse of the Antilles." *World Literature Today* 63, no. 4, Édouard Glissant Issue (1989) pp. 637–648.

———. "Beyond Miranda's Meanings: Un/Silencing the 'Demonic Ground' of Caliban's 'Woman'." In *Out of the Kumbla: Caribbean Women and Literature*. p. 93–98. Africa World Press, 1990.

———. "Unsettling the Coloniality of Being/Power/Truth/Freedom: Towards the Human, After Man, Its Overrepresentation—An Argument." *CR: The New Centennial Review* 3, no. 3, 'Coloniality's Persistence' (Fall 2003), pp. 257–337.

Chapter 3

Philosophizing Extinction

On the Loss of World and the Possibility of Rebirth through Languages of the Sea

Marjolein Oele

For my soul-sister Carole (1968–2019),
For presciently thinking through the dynamics of water

INTRODUCTION

My childhood is cast in the colors of the sea. There is the greyish-green of the IJssselmeer, the former Zuiderzee ("South Sea") that became a freshwater lake after it was dammed in 1933. Close to our family home, we swam here on hot summer days to cool off in its sweet-tasting waters. There is the greyish-blue color of the Oosterschelde—an estuary in Zeeland whose waters radically changed after sea barriers were placed as part of the Deltawerken-project in the province of Zeeland—where we returned each summer to the native grounds where both my parents were born and raised. I remember a day spent watching my mother there, the look into her past that place afforded:

> I watch closely as my mother and her oldest sister plan for their swim excursion. This must have been a daily part of their youth, of their sisterhood, of their family summer routine. Freedom found amid the waves. The joy and power of their young female bodies, so constrained in so many other ways.
>
> Their excitement at scanning the tidal schedules is palpable, contagious, even late in adulthood. It means so much more than just physical exercise. Their daily clocks lined up with the tides of the sea, they immerse and emerge. An oceanic rebirth?

The nostalgia palpable in this memory does not preclude the darkness that the same water holds. It was in these waters that close to 2,000 people, in the province of Zeeland alone, drowned in the North Sea Flood of 1953. The same flood also affected Belgium, England, and Scotland, causing casualties in these places as well.[1] My mother, who had just recently acquired her driver's license at the time, still trembled whenever she recounted chauffeuring people to safety in the days after the flood.

Thus, these transformed seawaters—the IJsselmeer and the Oosterschelde—inform my being with colors that go deeper than my own recollection and even that of my parents. Insofar as one is ever a citizen of a nation, the above stories about my time at sea evoke the way a country and a culture, in this case the Netherlands, has related to its surrounding seas. What reverberates with these stories about water are not only personal anecdotes, but stories of a country emerging out of the water, its vulnerability to water, and its consequent battle to control water. And in the particular case of the province of Zeeland,[2] what is important to recall is also its fraught attempts to subdue other parts of the globe with its seafaring skills, its horrendous trans-Atlantic slave trade,[3] and its colonial power-dynamics. If the sea is philosophically and historically characterized as delimiting the boundaries of the place we might call home, and houses not only dynamic fluidity but also openness and infinite possibilities, then the people of Zeeland radically transformed the relationship to their sense of home; they did so by altering their relationship to the sea, either through historical (colonial) globalizing navigation or through technological interventions such as dams and dikes, which changed the flow and character of the sea and changed the land as well. In both cases, the relationship to the sea became instrumentalized, transcending the liminal spaces that had offered a counterpoint of potentiality in order to actualize a particular passageway that would diminish the binding sense of belonging, of home, to the sea.

As I find myself on yet another shore, at another sea—now, the bright-blue Pacific Ocean, off the coast of Marin, with its cold, uninviting waters and treacherous currents—my life's relationship to the sea comes into focus. A perhaps somewhat ungrounded but fortunate migrant, I have had a hard time deciding what or who to call "home" but have settled on a preliminary answer, with home found not in the so-called solid substances, such as ground or nation, but in the connections, the shores, the transitions, and the interfaces to the open, in short, in the passageways that are made in and through the water.

If interaction with the sea is key to a sense of "home" on the planet—in my own autobiography as well as in the larger story of human life and all life as such—and if it is the case that our home is currently undergoing serious threats due to factors associated with industrialization and capitalism, then how should we grasp the loss associated with extinction and the loss that is emerging due to our changed relations with the sea? And what would this imply for the kind of responses we could offer?

This chapter's approach to the question of the loss of home and extinction will find its point of departure in a renewed understanding of the concepts of death and loss. Using the philosophy of Peter Sloterdijk, I will argue that the kind of losses we face vis-à-vis oceanic extinction—as part of the Sixth Great Extinction—are, in the first place, losses to be conceptualized not just on the level of the individual, but on the level of the constellation and the milieu. Secondly, I will argue that the narrative of death and extinction may be reframed as a narrative of revitalization, but only if we follow Michel Serres' lead in *Biogea*, where he seeks to open our senses and discourses to truly sensing loss so that we can co-emerge differently through a new semantics and epistemology that is truly responsive to the forces of life and Earth. Karin Ingersoll's proposal to offer an Indigenous seascape epistemology fills in some of Serres' ideas regarding formulating a language responsive to the sea, and aligns with Wesley Leonard's focus on language reclamation, since Ingersoll similarly seeks for a reclamation that, in Leonard's terms, should be "anchored in Indigenous needs and ways of knowing."[4]

Finally, Édouard Glissant's archipelagic thinking adds a much-needed explicit ethical and political layer to a discourse that is informed by the sea. Accordingly, I turn to the Caribbean islands and Glissant's work to argue that the experience of a shared loss as manifested on its shores and seas need not only be negative but may be generative in fostering an ethics and politics that is fundamentally organized in a multilinguistic and nonhierarchical fashion. I close by returning where I began: life and loss as lived on the increasingly less insular islands of Zeeland.

EXTINCTION AND LOSS OF WORLD AND HOME

With the question of extinction comes the question of the meaning of what is (or will be) lost. Usually, when we contemplate loss, we think of the loss of individualized beings, such as in Heidegger's famous account of being-toward-death, which focuses on the loss of a unique being confronted with its own finitude.[5] Peter Sloterdijk offers an alternative account of loss since he explores connections with others on both a micro- and on a macro-spherological levels. Accordingly, in *Bubbles*, Sloterdijk argues that life is always lived—on the intimate level—in biune spheres (orbs with differentiated, yet intimately connected spheres) bordered by semipermeable membranes that keep us both protected and open. And, as he addresses the macro-level, as our bonds grow deeper and older, and the membranes stretch thinner and more fragile, we can speak of an increased "shared existential risk."[6]

This furthers an account of death and loss that articulates we do not merely witness death in terms of individual losses, but that death implies the falling apart of a constellation, as well as having effects on all those involved in such a sphere. As Sloterdijk states:

> What Heidegger called being-toward-death means not so much the individual's long march into a final solitude anticipated with panic-stricken resolve; it is rather the circumstance that all individuals will one day leave the space in which they were allied with others in a current, strong relationship. That is why death ultimately concerns the survivors more than the deceased.[7]

Thus, with Sloterdijk, we find an opening to a more participatory account of death and loss, which emphasizes loss on the level of the milieu, and on the level of those who remain.

If we extend this analysis to the threat of extinction and oceanic changes near my childhood home and the North Sea, research indicates a shift in terms of cold-loving (plaice and flounder) to heat-loving (sea bass, shrimp) species.[8] What would happen to crustaceans[9] living in the North Sea and the Oosterschelde if the pH dips further and their shells dissolve?[10] What if we focus on one of the 52,000 species of crustaceans, for instance on the sea snail living in the North Sea shores called the common periwinkle (*Littorina littorea*)—in Zeelandic: "krukel"—and imagine its extinction? The common periwinkle lives in the intertidal zones, above water level at low tide, and underwater at high tide.[11] Should the periwinkle disappear, what would be lost is its connection to the tides, its immersion and emersion from the waters, its ability to experience both the wet and the dry, its liminal position on rocks and kelps along the seashore, and its way of grazing algae, not to speak of its specific construction of its own microclimate in the sea conducive to building its shell.[12]

On the human side, the krukel has its gastronomic appeal—common in Britain, France, Belgium, and part of the Netherlands—and this was also part of my family tradition: collecting, cooking, and eating them. Finally, should we lose the periwinkle, we may lose the connecting evolutionary bridge to land snails and its tapping into deep time, as "[i]t may be assumed that one of the paths leading to the evolution of terrestrial snails, went by the ancestors of both periwinkles and terrestrial operculate snails."[13] In sum, should the common periwinkle disappear, we lose with it its fluid, intertidal existence, holding steadily onto rocks while waves crash above, we lose the ability to close off or open up in the proper circumstances.[14] We lose its unique sense of being, and its impact on the deep and broad web of existence of which it has been part.

FROM A NARRATIVE OF DEATH AND EXTINCTION TOWARD A NARRATIVE OF AQUATIC REVITALIZATION

Amid the Great Sixth Extinction and the devastating impacts of climate change, Michel Serres in *Biogea* implores us to truly sense and grasp the loss that is taking place. Speaking to the sea—"the living sea, vital, vivifying, first mother of the living species"—means we have to (a) acknowledge the violence being done to her: "the sea soiled by hoodlums, violators of their first mother."[15] For Serres, the acknowledgment of loss needs to be accompanied by examining its causes. However, asserting and examining loss is not enough. Serres aims for us to move through and beyond loss toward a possible rebirth of our Earth and its diverse species. To do so, Serres provides us with (b) a distinct ontology that views all forms of existence (both organic and non-organic) as interconnected and communicating. The task he outlines next is that of (c) generating a new semantic realm that taps into this communicating ("coding") realm of reality with more receptive, holistic connections.

(a) As Serres is seeking a culprit at the heart of the pollution and extinction that threatens nature, he finds not one in particular, but the murderous capacity of a multitude: "persons sometimes kill; the collective always kills."[16] It is the collective that is to blame,[17] who names and throws objects before us and considers them rejected and disposable: "trashcan-Earth, polluted air, dead seas . . . soiled by us for us to appropriate them. Destroyed by a collective that's narcissistic in its turn."[18] And in analyzing the various steps in history that have brought us to the point of collapse, Serres does not shy away from pointing to brilliant scientists in Western history, such as Archimedes, to show how Western thinking,[19] from the beginning, has been preoccupied with competition and warfare.[20]

Additionally, Serres blames disrespect for Earth and life on the current dominant European forms of language,[21] which has put "all the things of the world [. . .] locked up behind their bars,"[22] removing us from having true contact with the phenomena themselves, and ossifying them in their stilted, objectifying approach. This approach has obscured reality for what it is, by accessing nature in a merely appropriative, anthropocentric, controlling, and calculative way:

> For thousands of years, we have been licking things with our tongues, covering and daubing them so as to appropriate things for ourselves. If language boils

> down to a convention, this convention took place between the speakers without consulting the thing named, become as a result the property of those who covered it in this way with their drawn or voiced productions.[23]

Instead, Serres seeks to engage a language that has "reciprocity,"[24] which means "to be able to understand what the earth is telling us but also effectively communicate for the earth the needs of all the animals and the planet itself."[25]

(b) The ontological view that Serres provides us of the world and nature is one of complexity, fluidity, chance, interconnection and communication:

> Caused, causing, all things in the world ensue from each other, chained together. Whether fluid or air—even solids communicate—things respire together, they conspire with different breaths, but in a constant and total circulation that's chancy, torn, chaotic and consenting.[26]

This interconnection is deep and variable, as this example indicates: "El Niño spreads ocean and atmospheric currents, brush fires, thunderstorms and draughts around the world."[27] In this holistic view of the Biogea, Serres distinguishes certain patterns—"music, waves, codes,"[28] while also emphasizing the role of chance: "Everything, on the contrary, in the unpredictable, via contingent lightning bolts and impacts. Everything via the diagonal, the oblique, the inclined, the random without reason."[29]

According to Serres, living and non-living things express themselves and do so in reciprocal interaction with each other; for instance, the howling of the wind happens when two different currents of wind—broken up by objects such as trees—mix together. The emergence of expression is, for Serres, born out of commotion, out of turbulence, and the same can be said about our own human language: that it is born out of movement, out of embodied e-motion, and expresses itself in certain kinds of rhythms or patterns.[30] Fundamental to this idea is that, for Serres, all (living and non-living things) "receive, transmit, store and process information."[31] Expressive content thus occurs beyond the (traditionally human) linguistic levels; it happens at all inorganic and organic levels.

(c) In the face of severe loss, Serres' interconnected viewpoint of the world demands a response that recognizes loss and seeks to install an ethical and political response that keeps open the future for generations to come. As the Biogea and its seas, rivers, and species suffer, it seems no one worries or cares about hearing "the death pangs of the rivers."[32] By truly grasping the changes that have occurred due to an extractivist human interference (for instance in the form of dams) and thinking through their consequences, can we also grasp the need for political action, so that the availability of drinking water for future generations may be guaranteed? [33]

To seek rebirth amid disaster, Serres invites us "to see like the sea, dying, or the rivers, weakened," like Aldo Leopold's invocation to "think like a mountain,"[34] but even more broadly so, as Serres' invocation functions by inviting us to sense and think "like the entire Biogea, the entirety of the Earth and the living species."[35] To think like the Biogea, we have to let go of simply connecting to the Biogea as if it were background noise to civilization; it has to emerge as the front and center of our lives. "As we recount encounters in the flux spaces of coding that Serres outlines, we provide expressive content to a new semantics."[36] However, for Serres, expressive content is not localized in any one arena but can find shape in science, coding, poetry, philosophy, music, sensation, embodiment, dreams, mythologies, and stories. According to Serres, some of these access paths have been walked by previous cultures: for instance, the Druid practice that saw human and tree life deeply interwoven and involved sacred ceremonies in forests honoring our mutual connectedness. And, adding on to Serres, it is important to note that some contemporary Indigenous people continue to practice lifeways like these. While none of these arenas alone may channel a rebirth, together such various access paths may allow us channels to holistically hear the voices of Earth and life.

For Serres, the importance of language in this endeavor is crucial. Not only does Serres here invoke us to move forward with a new discourse, he also reflects on now "silent" words that spoke to a deep connection with the elements, as well as to a community.[37] For instance, as Serres reminisces on the dialect of his youth and the specific words connected to the barge trade of his father on the river Garonne, he writes: "but I bemoan the silence of some fifty words only used with ten bargemen, my father and brother, all vanished."[38] The richness of this now "mute language" speaks to the complexity and the comradery of life lived on the platoon boat, keeping it and each other safe amid the dangerous currents and floods of the river.

In reference to his youth, Serres shows how certain languages and/or dialects may be better attuned to the vicissitudes of the real. At the same time, Serres also points at a contradiction: the endless multiplication of dialects and languages that have made humans so separate from each other. For Serres, a new language, one receptive to and born out of the Biogea, will need to be conceived. This is a "soft" place as opposed to the "hard," where Serres associates soft with "culture, concepts, ideas, sign, meaning, and, generally, human accounts of reality," and hard with "the domain of nature, the given, the physical, the world of objects, reality independent of human perception of it."[39] He continues:

> Without this soft place, spiritually very old, but newly conceived in this way, without the juridical construction of a common good, opposed to our

> filthy ownership, I don't see how our planet, hard, will survive. Hardness that depends on softness, material belonging that depends on this temporary rented location.[40]

For Serres, the survival of our world—the "hard" matter of the Biogea—will depend on the new "soft" construction of meaning and language, along with a new politics, law, and so on. While I would like to criticize Serres for drawing an opposition between "hard" and "soft" which, at least at first sight, seems to lead us back to problematic dualisms such as nature vs. culture, or fact versus value, Serres' intention may be appreciated in its radicality: he thinks that without a new epistemological, linguistic, and social paradigm, not only we, humans, but the entire Earth is prone to decay. And, by highlighting the importance of the "common good," Serres also clearly indexes how by tapping into the "universal language" of the Biogea a new language should steer us in our new conceptions: from individuality to commonality, from ownership to temporary tenancy, from separation to interconnection.[41] To exemplify what the grammar of such new languages should entail, Serres highlights his "philosophy of prepositions" which prioritizes random encounters that allow for co-birth and rebirth with other beings[42] and marginalizes focus on subjects and verbs which signify a solitary existence.[43]

THE LANGUAGE OF THE SEA, THE DUTCH CARIBBEAN, AND GLISSANT

If we should give heed to Serres' quest for a language that is responsive to the Biogea, and one that is informed by "the codes" emitted by living beings and things, what would a language be that is responsive to the sea, and what might guide the creation of such a language?[44] Who or what may guide the way, and who or what may consequently express its danger, its current vulnerability to extinction, but also its ability to be reborn?

Following Karin Amimoto Ingersoll's book *Waves of Knowing*, we could argue that the sea offers us what Ingersoll speaks as a "seascape epistemology,"[45] which is grounded on "local ways of knowing, researching, and producing knowledge"[46] as lived and embodied by Indigenous Hawaiians (Kãnaka Maoli) rather than as imposed by a colonial reality.[47]

For her, seascape epistemology

> is an approach to knowing through a visual, spiritual, intellectual, and embodied literacy of the *'āina* (land) and *kai* (sea): birds, the colors of clouds, the flows of the current fish and seaweed, the timing of ocean swells, depths, tides, and celestial bodies all circulating and flowing with rhythms and pulsations [. . .]

> Seascape epistemology embraces an oceanic literacy that can articulate the potential for travel and discovery, for a re-creation and a de-creation.[48]

Ingersoll's focus on epistemology as based on relationships to land and sea echoes Leonard's claim about the reclamation of Miami language; namely, that such reclamation is ultimately about restoring relationships.[49] Both Ingersoll and Leonard emphasize that oceanic knowledge (Ingersoll) or the reclamation of language (Leonard) depends on what Leonard references as the "physical, spiritual, and relational contexts"[50] that underpin such knowledge or language.

The advantage of Ingersoll's account over Serres' is that it highlights a particular phenomenology of the sea through a non-Western, Indigenous perspective. And, in contrast to Serres' personal anecdotes of his time in the navy or the male comradery of the pontoon boat, Ingersoll offers a view on a distinctly inclusive and collaborative Hawaiian collectivity living with and off the sea. And, while Ingersoll's account initially seems restrictive, compared to Serres' more global call, it is arguably the case that (re)vitalizing local knowledges and practices are a way toward regenerative (global) futures.

Another track for an aquatic language may follow the currents of a troubled colonial Dutch past to the Caribbean. The Caribbean islands include the Dutch Caribbean islands located in the "Lesser Antilles archipelago of the Caribbean Sea,"[51] which were used by the Dutch in the eighteenth century as "depots" or portals for the transatlantic slave trade.[52] To think through a language of the sea that speaks to this archipelago informed by colonialism, Édouard Glissant's *Poetics of Relation*[53] will be particularly productive.

Glissant shows us the profound meaning of this sea-informed place as well as the repercussions the slave trade had on those who arrived on the islands. The harrowing experience with which Glissant's *Poetics of Relations* begins is that of being cast into the belly of a boat, of seeing fellow humans being thrown overboard as heavy ballast to lighten the boat, and arriving upon a new unknown land, haunted by the mourning of the former land. He writes:

> For though this experience made you, original victim floating toward the sea's abysses, an exception, it became something shared and made us, the descendants, one people among others. Peoples do not live on exception. Relation is not made up of things that are foreign but of shared knowledge. This experience of the abyss can now be said to be the best element of exchange.[54]

Rather than seeing the experience of the sea's abyss only from the perspective of loss or from the perspective of the singular individual, Glissant demonstrates that a new future may be built upon a collective memory and a shared knowledge. Politically, this may allow for hope to emerge out of suffering and may provide a vision of the possibility of an epistemology that is born out of multiplicity, yet built out of shared knowledge: an epistemology of relations.

Glissant's embrace of a vision of synthetic possibility as emerging out of the Caribbean stands in contrast to two other influential Caribbean thinkers: Césaire and Fanon. "For Glissant, both Césaire and Fanon are still diverters and not properly producers of a new reality, of a real Caribbean territory and history." While Glissant finds himself located in-between those two thinkers, Glissant

> neither advocates a reconstruction that points to a past located elsewhere (Africa), nor recommends the rejection and replacement of the here/now with a different, unknown spatiality and temporality. For Glissant, the locus of resistance is located in the present and in the possibilities of decolonization already contained in the Caribbean, although concealed and understated.[55]

Thus, rather than simply yearning for a lost past or turning to a future-yet-to-come, Glissant finds a synthetic notion of possibility emerging from the here and now of the Caribbean.

Moreover, as Ramírez argues, the poet and thinker Césaire expressed

> contempt for the Caribbean landscape: the beach (and the sea), the island and the archipelago. In all three, the poet Césaire suggests the complicity of space in the people's muteness and inert character and thus the landscape itself becomes an ally of the oppressor and not a tool for liberation and resistance.[56]

By contrast, Glissant embraces the opportunity of the very locality of the Caribbean and its sea-infused, archipelagic existence.

Additionally, and most prominent for my argument here, Glissant argues that the Caribbean formation of language (which also has had traces of a former Dutch, and even Zealandic past)[57] offers, through its openness, a decentered and multilingual way to think and live. The genius of the Creole language consists for Glissant:

> [. . .] in always being open, that is, perhaps, never becoming fixed except according to system of variables that we have to imagine as much as define. Creolization carries along then into the adventure of multilingualism and into the incredible explosion of cultures.[58]

Rather than synthesizing differences into a concentrated unity, the power of the multilingualism that Glissant addresses lies in its ability to multiply, to be organized nonhierarchically, and to vary time and again.

Thus, aquatic speech acts such as those that emerge in the Creole language show that a new language of the sea need not be romanticized (as is possibly suggested by Serres) or be referenced solely through traditional ways of being (such as the native Hawaiian way-of-living-and-relating as is articulated in Ingersoll). Rather, with Glissant, we come to recognize that a language of the sea may evolve geographically-naturally-culturally out of a variety of stances, out of an erratic mixture of languages, and thus may carry with it horizontal power-structures extending into the future, with no one language being ascertained as the "center" of hierarchy. Creolization, as Drabinksi clarifies, is "a passage that works *against* conventional models of transition. Diffraction distorts, bends, and chaotically reconfigures."[59] Accordingly, the diffraction of which Glissant speaks may add a helpful caveat insofar as any imagination of a (future) language of the sea should not be monolithic nor hierarchically organized.

To exemplify Glissant's thinking on languages of the sea further, the tradition of "tambú" as practiced on the Dutch Caribbean island of Curaçao is apt to discuss.[60] With origins in West Africa, tambú includes a drum (tambú) or barí (barrel), at least two iron instruments (known collectively as herú), vocals and dance.[61] Originating in the seventeenth and eighteenth centuries, the vocals are spoken in Papiamento, which is comparable to the Creole language of which Glissant speaks, a mixture that has evolved, blending various African languages and Portuguese with aspects of Spanish, English, and Dutch.[62]

Much like the synthetic possibilities that Glissant indexed, the tradition of tambú stands at a critical distance from events and urges life to go on, but differently, more perceptively, more critically, and with improvization. Tambú parties, which combine music and dance with food and drinking, offer an additional element of social cohesion and solidarity. Ranging from resistance to critical commentary, tambú signals the power of a musical language that drives life lived on the islands further in a generative direction.

Given its reflective possibilities, it is interesting that many Curaçaoans have acquired negative connotations with Tambú parties, possibly due to the fact that the parties associated with Tambú attract rebellious youth, heavy drinking, and sometimes result in acts of crime. However, for those Curaçaoans living in the Netherlands, such parties are looked upon much more favorably.[63] In fact, given that

> Tambú had a turbulent past, scrutinized by both the governing Dutch and local Curaçaoans, the parties enabled Curaçaoans in the Netherlands the opportunity not only to rebel against the Netherlands establishment, but to break with local mores on Curaçao as well. As a result, Tambú parties evolved into a strategy of resistance, where cultural expectation and cultural authority could be renegotiated.[64]

Provocatively, Tambú parties in the Netherlands afford not only displaced Curaçaoans but many migrants from West and North Africa and the Caribbean as such the opportunity to find what De Jong calls a "space for collective belonging,"[65] where each individual may insert their own identity through dance moves reflecting cultural sensibilities and potencies.[66] Rather than nostalgic belonging for a home left behind, they are capable of doing what Glissant has argued for: a synthetic sense of generating home out of loss and displacement and doing so through a multilinguistic and nonhierarchical sense of music and dance.

LANGUAGES OF THE SEA

As I return to the issue of loss with reference to the sea-born islands of Zeeland, a complex, paradoxical situation presents itself. On the one hand, due to the construction of the "Deltawerken"—an elaborate system of dams, barriers, and dikes which connect the islands to each other and the mainland—Zeeland's population has benefited from increased safety, prosperity, healthcare, and so on. Also, due to increased individualization of its population and the fraying of local municipal bonds (originally infused by religious differences and rivalry between municipalities), collaboration across municipalities has increased, as well as cohesion with the Netherlands and Belgium.[67]

On the other hand, decisions regarding the ongoing sea risks that threaten Zeeland, which have to do with rising sea levels and the risk of flood, have increasingly made the population of Zeeland less resilient: as the population progressively lives more removed from the sea, and as a brain-drain removed its sea-knowledgeable citizens from its rural areas to other locations in the Netherlands, water management became increasingly centrally coordinated from elsewhere, and accordingly, the population became increasingly removed from decision-making processes. Along with the issue of human resilience is the issue of the ecosystem. Even if the ecological effects of its dams and dikes are currently under discussion and prone to changes, the net effect of the Deltawerken on the ecosystem remains astounding. If we take a look at the estuary of the Oosterschelde, once called the "incubator of the North Sea," one Zeeland skipper declares: "The water is simply dead."[68] As Zeeland's sea-and-rural landscape has become homogenized and emptied out, the sense of "home"—a deep connection to the shores, transitions, and the interfaces to the open—seems lost. How can this sense of loss and estrangement from water transform into a renewed sense of belonging, a new-found resilience?

If it is the case, as Wittgenstein argues, that "to imagine a language is to imagine a form of life,"[69] it is the task for philosophy to think through possibilities of a new language—or better said: new languages—that are both informed by the loss of the world as world, but also trusting that imaginative forms of life and culture may emerge or be sustained. In moving forward, the model of the "seagoing pact" as Serres relates it may serve us well. Notably, this is a different kind of collective than the one addressed earlier, which was an objectifying, possessive "we." The seagoing pact, instead, is one that co-emerges with the elements, in symbiosis. Much like the pontoon boat in the Garonne, here "the collectivity, if sundered, immediately exposes itself to the destruction of its fragile niche."[70]

Thus, the notion of loss has circumnavigated the globe. Where do we go from here? As we navigate the troubled waters of ecological destruction, the image of collective navigation and of seeking relation-in-difference may guide us, I hope, to creatively and peacefully connect to safer shores.[71]

NOTES

1. "The 1953 North Sea flood was a major flood caused by a heavy storm that occurred on the night of Saturday, 31 January 1953 and morning of Sunday, 1 February 1953. The floods struck the Netherlands, Belgium, England and Scotland. A combination of a high spring tide and a severe European windstorm over the North Sea caused a storm tide; the combination of wind, high tide, and low pressure led to a water level of more than 5.6 meters (18.4 ft) above mean sea level in some locations." The storm caused great damage to the coastline in both England and Scotland, breaching seawalls and inundating large areas. Wikipedia, s.v. "North Sea Flood of 1953," accessed September 4, 2019, https://en.wikipedia.org/wiki/North_Sea_flood_of_1953. Cf also: D. Rollema, "Amateur Radio Emergency Network During 1953 Flood." *Proceedings of the IEEE* 92, no. 4 (April 2004): 759–62, https://ieeexplore.ieee.org/document/1278696.

2. The name of Zeeland can be found in various places around the world connected with its colonial history, such as in the names of four fortresses, all called "Fort Zeelandia": *Fort Zeelandia*, in current day Taiwan, in Suriname, in Guyana, and in Benin. Wikipedia, "Fort Zeelandia (Taiwan)," accessed May 20, 2020, https://en.wikipedia.org/wiki/Fort_Zeelandia. Cf. also Gert Costindie, "Index of Geographical Names," in *Dutch Colonialism, Migration and Cultural Heritage* (Leiden, The Netherlands: Brill, 2008), https://doi.org/10.1163/9789004253889_009.

3. As discussed in Ryan Heryford's chapter for this volume, a point in case is the slave ship *Zong*, originally called *Zorg*, which means paradoxically and cruelly, in Dutch, "care." While the case Heryford discusses pertains to the genocide that took place under British ownership of the slave ship, the ship itself originally was a Dutch boat, owned by the Middelburgsche Commercie Compagnie, based in Middelburg, in what is now the province of Zeeland in the Netherlands. In 1777, it delivered enslaved people to Suriname and the ship was later captured by the British. Jane Webster, "The *Zong* in the Context of the Eighteenth-Century Slave Trade," *Journal*

of Legal History 28, no. 3 (November 2007): 288, doi:10.1080/01440360701698403. Reference is to Heryford in this volume, chapter 2, page 70.

4. In his chapter "Contesting *extinction* through a praxis of language reclamation," Wesley Leonard clarifies the importance of restoring language anchored in Indigenous needs and ways of knowing. He differentiates this from language "revitalization" projects which follow dominant notions that language is to be "fixed." As should become clear in this chapter, my own use of the term "revitalization" is more closely aligned with Leonard's term "restoration" than with what he denotes as the problematic practice of language "revitalization." Reference is to Leonard in this volume, chapter 6, pages 145–146.

5. Heidegger's account of being-towards-death has been critiqued from multiple angles, for instance by Edith Stein. Stein shows, versus Heidegger, the importance of socialization for our sense of death, and how experiencing the death of others is crucial in relationship to our own understanding of death. Edith Stein, "Martin Heidegger's Existential Philosophy," trans. Mette LeBech, *Maynooth Philosophical Papers* IV (2007), 55–98, at 76–7.

6. Peter Sloterdijk, *Foams: Spheres Volume III: Plural Spherology*, trans. Wieland Hoban (South Pasadena: Semiotext(e), 2016), 48.

7. According to Sloterdijk, there is however not only dissolution, but also the possibility of sublation: "Human death thus always has two faces: one that leaves behind a rigid body and one that shows sphere residues—those that are sublated into higher spaces and re-animated and those that, as the waste products of things, fallen out of former spaces of animation, are left lying there" (Sloterdijk, *Foams*, 48).

8. "It is likely that, with a further rise in temperature, the North Sea will become more livable for southern species and that the number of northern species will decrease. Climate change can also have indirect effects on species by acting on the food web. Because of warming up, seawater can contain less oxygen, which is unfavorable for certain species" (my translation). Ralf van Hal, Oscar Bos, and Robert Jak, "Noordzee: systeemdynamiek, klimaatverandering, natuurtypen en benthos; Achtergronddocument bij Natuurverkenning 2011," Wageningen, Wettelijke Onderzoekstaken Natuur & Milieu, WOt-werkdocument 255, at 9, https://library.wur.nl/WebQuery/wurpubs/fulltext/189129.

9. Crustaceans include commonly-known marine life such as crabs, lobsters, barnacles, and shrimps [. . .] there are over 52,000 species of crustaceans." Jennifer Kennedy, "Crustaceans: Species, Characteristics and Diet," ThoughtCo, last updated December 13, 2019, https://www.thoughtco.com/crustaceans-profile-and-facts-2291816.

10. As the pH drops, "[c]rustaceans, such as shellfish, may be affected by this in the future because their shell can dissolve" (my translation; Bos, van Hal, and Jak, "Noordzee," 9).

11. Wikipedia, s.v. "Intertidal Zone," accessed August 21, 2019, https://en.wikipedia.org/wiki/Intertidal_zone. Cf. also: Samuel Kinuthia, "What is the Intertidal Zone?" Environment, World Atlas, January 10, 2018, https://www.worldatlas.com/articles/what-is-the-intertidal-zone.html.

12. Elizabeth Kolbert, *The Sixth Great Extinction: An Unnatural History* (New York: Simon & Schuster, 2014), 121.

13. Robert Nordsieck, “Amazing Facts About Snails,” in The Living World of Molluscs, accessed August 31, 2019, http://molluscs.at/gastropoda/index.html?/gastropoda/sea.html.

14. Nordsieck, “Amazing Facts About Snails.”

15. Michel Serres, *Biogea*, trans. Randolph Burks (Minneapolis: Univocal, 2012), 10.

16. Serres, *Biogea*, 17.

17. Serres, *Biogea*, 22.

18. Serres, *Biogea*, 32. Additionally, speaking to so-called “natural disasters,” Serres points out that it is problematic “political, economic, social conditions—poverty for example—[that] prevail” in exacerbating death-counts, thus clarifying that the suffering experienced is unequally and dramatically experienced most by those who are marginalized (Serres, *Biogea*, 29).

19. Serres’ way of framing the collective in this passage, namely in terms of Western history and science, indicates that he is not blaming the entirety of humanity for pollution and extinction, but the destructive and polluting Western systems of thought that humans participate in. For more examples of the destructive power of such systems, see the chapter of Figueroa, Perez and Mantz which analyzes how, in contrast to Indigenous practices, “colonial, Eurocentric, patriarchal and capitalist industrial civilization has pushed the planet to mass extinction, biocultural obliteration and climate catastrophe in a few hundred years” Reference is to Figueroa, Perez and Mantz in this volume, chapter 1, page 28.

20. Serres, *Biogea*, 67. In the case of Archimedes, Serres discusses his brilliance in math and physics, as well as his involvement in warfare, setting up gigantic mirrors in the landscape that, when attracting sunlight, would let the ships of the Roman enemy fleet ablaze.

21. For instance, Leonard, in his chapter for this volume, speaks of the tendency of English to be noun-based, thereby omitting other conceptions such as that language could also be a verb, as it is for the Miamis who consider their own term for language as a verb: “myaamia is spoken.” Reference is to Wesley Y. Leonard in this volume, chapter 6, page 152.

22. Serres, *Biogea*, 38.

23. Serres, *Biogea*, 38–39.

24. A similar focus on reciprocity is discussed by Figueroa et al., for instance here: “For Indigenous peoples, interpersonal relational intimacy weaves humans and the land into integrated communities where the holistic regeneration of life depends on reciprocal human-nonhuman nurturance.” Reference is to Figueroa, Perez and Mantz in this volume, chapter 1, page 29.

25. Brandon Guerrero, *Hell is a Place on Earth: USF Existentialism Paper* (San Francisco, CA: May 15, 2020), 1.

26. Serres, *Biogea*, 129.

27. Serres, *Biogea*, 128.

28. Serres, *Biogea*, 129.

29. Serres, *Biogea*, 154.

30. More than simple interactions, much more is at stake. Serres points at orchestral music to capture the emergence of rhythms and patterns in the world. However, pattern here is not all there is: sound is accompanied by background noise, order is bound by disorder (Serres, *Biogea,* 111).

31. Serres, *Biogea*, 196.

32. Serres, *Biogea*, 24.

33. Serres, *Biogea*, 24.

34. Aldo Leopold, *A Sand County Almanac, and Sketches Here and There* (Oxford: Oxford University Press, 1989), 113.

35. Serres, *Biogea*, 23.

36. Daniel O'Connell, email correspondence with the author, May 21, 2020.

37. Cf. Leonard on how reclamation praxis is ultimately a story of relationships: in the myaamia concept of language, "Miami people and our relational interactions [are embedded] into the definition" of language. Wesley Y. Leonard in this volume, chapter 6, page 152.

38. Serres, *Biogea*, 12.

39. Brian Treanor, "Mind the Gap: The Challenge of Matter," in *Carnal Hermeneutics*, ed. Brian Treanor and Richard Kearney (New York: Fordham University Press, 2015), 67.

40. Serres, *Biogea*, 51.

41. Serres emphasizes the importance of *connecting* various kinds of knowledge, to return to the inner *combination* that is part of the things of the world as well. This new, synthetic kind of knowledge is characterized by "sets united by interlacings, webs and simplexes that combine with the things of the world, themselves combined, the combined knowledge that understands them" (Serres, *Biogea*, 131).

42. Similar to how Serres highlights prepositions, Leonard in his chapter emphasizes how for the Miami community, language is defined in terms of *interactions* (Leonard, 152) with special emphasis on strong relationships as expressed by what are called "'the R-words' such as respect, reciprocity, and responsibility" that are key to the notion of strong relationship that stands at the core of Miami culture (Leonard, 153).

43. Cf. also Serres' "philosophy of prepositions" as discussed in: Michel Serres with Bruno Latour, *Conversations on Science, Culture, and Time*, trans. Roxanne Lapidus (Ann Arbor, MI: The University of Michigan Press, 1995), 103. Serres wants to turn away from a thinking and language that is centered around substantives and verbs, to a language of prepositions which emphasizes relations. Serres writes: "To talk only by means of substantives or verbs, and thus to write in a telegraphic code, as ordinary philosophy does, defines a different form of abstraction from the one I propose, which relies on prepositions" (103).

44. Posthumus argues that, for Serres, "language does not remove us from the material world; it is part of this world, as the water in which thinking swims. Ecological dwelling thus necessarily includes the practices of language." Stephanie Posthumus, *French Écocritique: Reading Contemporary French Theory and Fiction Ecologically* (Toronto: University of Toronto Press, 2017), 67.

45. Karin Amimoto Ingersoll, *Waves of Knowing: A Seascape Epistemology* (Durham-London: Duke University Press, 2016), 5.

46. Ingersoll, *Waves of Knowing*, 3.

47. Ingersoll, *Waves of Knowing*, 6.

48. Ingersoll, *Waves of Knowing*, 6.

49. Leonard, chapter 6, 153.

50. Leonard, chapter 6, 150.

51. The Dutch Caribbean includes the islands of Bonaire, Saint Eustatius and Saba (also known as the "Caribbean Netherlands") and the islands of Curaçao, Bonaire and St. Maarten (which are part of the Kingdom of the Netherlands). "What are the different parts of the Kingdom of the Netherlands?" Question and Answer, Government of the Netherlands, accessed May 20, 2021, https://www.government.nl/topics/caribbean-parts-of-the-kingdom/question-and-answer/what-are-the-different-parts-of-the-kingdom-of-the-netherlands.

52. Cf. the analysis of the transatlantic slave trade as discussed in Ryan Heryford's chapter for this volume. Reference is to Heryford, chapter 2 in this volume, 71.

53. Édouard Glissant, *Poetics of Relation*, trans. Betsy Wing (Ann Arbor: The University of Michigan Press, 1997). He writes: "Without necessarily inferring any advantage whatsoever to their situation, the reality of archipelagos in the Caribbean or the Pacific provides a natural illustration of the thought of Relation" (33).

54. Glissant, *Poetics of Relation*, 8.

55. Miguel Gualdrón Ramírez, "To 'Stay Where You Are' as a Decolonial Gesture: Glissant's Philosophy of Antillean Space in the Context of Césaire and Fanon Chapter," in *Memory, Migration and (De)Colonisation in the Caribbean and Beyond*, eds. Jack Webb et al. (London: University of London Press; Institute of Latin American Studies, 2020), 134, https://www.jstor.org/stable/j.ctvwh8cwp.14.

56. For instance, as Ramírez explains, the beach and waves crashing on the shore signifies the draining of a life force being sucked away, the arrival of death and decomposition (Ramírez, "To 'Stay Where You Are' as a Decolonial Gesture," 138).

57. Cf. Jacques Van Keymeulen, "Zeeuws Overzee," in *NEHALENNIA* 198 (2017): 18–19, who reports on a Creole language based on Zeelandic which is known as "Negerhollands" (negroDutch) that used to be spoken on the Danish Antilles (Saint-Thomas, Saint-Croix and Saint-John). "[The Danish Antilles] were sold by Denmark to the United States in 1917 and are currently called the American Virgin Island (Virgin Islands of the United States). When in 1666 the British attacked the Dutch Island St. Eustatius, numerous Zeelandic and West-Flemish 'planters' [trl?] fled to the Danish island and their slaves brought the creole Dutch with them as 'plantation language.' White people were also able to speak the language. The last speaker of the mother tongue of Negerhollands, Alice Stevens, died in 1987" (Van Keymeulen, 2–3).

58. Glissant, *Poetics of Relation*, 34.

59. Drabinski contrasts diffraction helpfully with transmission, which "*promises* (even if it cannot realize) the preservation of what is prior in what is later" (his italics). John E. Drabinski, *Glissant and the Middle Passage: Philosophy, Beginning, Abyss* (Minneapolis: University of Minnesota Press, 2019), 82.

60. “Curaçao, an island of harbors,” Curaçao Maritime Museum, accessed May 11, 2020, https://www.curacaomaritime.com/history.

61. Only since 1952 has the tambú become legal, after “200 years of condemning, condoning and forbidding.” Interestingly, tambú parties are only allowed for a 3-month period per year, known as the “tambú season.” In 2015, tambú was placed on the National Inventory of Intangible Cultural Heritage in the Netherlands on the initiative of the SPLIKA Foundation. Splika Foundation Online, “Ontdek Tambú!” § 1–5, accessed May 11, 2020, https://splika.nl/ontdek-tambu/.” See also Nanette de Jong, “Curaçao and the Folding Diaspora: Contesting the Party Tambú in the Netherlands.” *Black Music Research Journal* 32, no. 2 (2012): 69, accessed May 22, 2020. doi:10.5406/blacmusiresej.32.2.0067.

62. Splika Foundation, “Ontdek Tambú,” § 2.

63. De Jong, “Curaçao and the Folding Diaspora,” 76.

64. De Jong, “Curaçao and the Folding Diaspora,” 74.

65. De Jong, “Curaçao and the Folding Diaspora,” 77.

66. De Jong, “Curaçao and the Folding Diaspora,” 77.

67. Emmen, *De Kwetsbaarheid Van De* Zeeuwen, online publication, 2002, 36–7, accessed May 16, 2020, https://www.ifv.nl/kennisplein/Documents/emmen-4-de-kwetsbaarheid-van-de-zeeuwen.pdf.

68. Interview with skipper Jan Molenaar, “Onder water zit het niet goed in de Oosterschelde,” interview by Totalfishing, accessed May 16, 2020, web, http://www.totalfishing.nl/laatste-nieuws/articles/onder-water-zit-het-niet-goed-in-de-oosterschelde.

69. Ludwig Wittgenstein, *Philosophical Investigations*, trans. G. E. M. Anscombe (Oxford: Blackwell, 1953), § 19/11e.

70. Michel Serres, *The Natural Contract*, trans. Elizabeth MacArthur and William Paulson (Ann Arbor: University of Michigan Press, 2011), 40.

71. In its various iterations, this chapter benefited from conversations with my USF colleagues Kim Carfore, Gerard Kuperus, Anne Mairesse, Sam Mickey, Amanda Parris and Ron Sundstrom and from the feedback of the participants of the 11th Annual Meeting of PACT “The Sea” in September 2019, and especially the comments from Russel Duvernoy, Michael Eng, Robert Mugerauer, Michael Shaw, Peter Steeves, Tom Thorp, Brian Treanor, Jason Winfree and Jason Wirth. I owe particular gratitude to Daniel O’Connell for his inspiring, reflective, and constructive feedback rethinking the various drafts of this paper. I also would like to thank my research assistants Darcy Allred and Lincoln Stefanello for their careful editorial and research assistance. I am tremendously grateful for the opportunity to have presented an earlier draft of this paper at the symposium Approaching Extinction | Contesting Extinction at Miami University and for all the critical and constructive feedback I received from the organizers and participants.

BIBLIOGRAPHY

Bos, Oscar G., Ralf van Hal, and Robert Jak. “Noordzee: systeemdynamiek, klimaatverandering, natuurtypen en benthos; Achtergronddocument bij Natuurverkenning

2011." Wageningen, Wettelijke Onderzoekstaken Natuur & Milieu, WOt-werkdocument 255. Wageningen: Wageningen University and Research, 2011. https://library.wur.nl/WebQuery/wurpubs/fulltext/189129.

De Jong, Nanette. "Curaçao and the Folding Diaspora: Contesting the Party Tambú in the Netherlands." *Black Music Research Journal* 32, no. 2 (2012): 67–81. Accessed May 22, 2020. https://doi.org/10.5406/blacmusiresej.32.2.0067.

Drabinski, John E. *Glissant and the Middle Passage: Philosophy, Beginning, Abyss*. Minneapolis: University of Minnesota Press, 2019.

Emmen, Ton. "The Vulnerability of the Zeelanders: Factors which Influence the Fragility and Resilience of the Population in Regard to the Threat of Floods from Sea." (Dutch: *De Kwetsbaarheid Van De Zeeuwen: Factoren die de kwetsbaarheid en veerkracht van de bevolking ten aanzien van dreiging van overstromingen vanuit zee beïvloeden*). Online publication, 2002. Accessed May 16, 2020. https://www.ifv.nl/kennisplein/Documents/emmen-4-de-kwetsbaarheid-van-de-zeeuwen.pdf.

Figueroa, et al. bibliographic information from this volume (chapter 1).

Glissant, Édouard. *Poetics of Relation*. Translated by Betsy Wing. Ann Arbor: Univ. of Michigan Press, 1997.

Guerrero, Brandon. *Hell is a Place on Earth: USF Existentialism Paper*. San Francisco, CA. May 15, 2020.

Heryford, Ryan, in this volume (chapter 2).

Ingersoll, Karin Amimoto. *Waves of Knowing: A Seascape Epistemology*. Durham and London: Duke University Press, 2016.

Kennedy, Jennifer. "Crustaceans: Species, Characteristics and Diet." ThoughtCo. Last updated December 13, 2019. https://www.thoughtco.com/crustaceans-profile-and-facts-2291816.

Kolbert, Elizabeth. *The Sixth Extinction: An Unnatural History*. New York: Simon & Schuster, 2014.

Leonard, Wesley Y., in this volume (chapter 6).

Leopold, Aldo. *A Sand County Almanac, and Sketches Here and There*. Oxford: Oxford University Press, 1989.

Molenaar, Jan. "Onder water zit het niet goed in de Oosterschelde." Interview by Totalfishing. Accessed May 16, 2020. http://www.totalfishing.nl/laatste-nieuws/articles/onder-water-zit-het-niet-goed-in-de-oosterschelde.

Nordsieck, Robert. "Amazing Facts About Snails." The Living World of Molluscs. Accessed August 31, 2019. http://molluscs.at/gastropoda/index.html?/gastropoda/sea.html.

Posthumus, Stephanie. *French Écocritique: Reading Contemporary French Theory and Fiction Ecologically*. Toronto: University of Toronto Press, 2017.

Ramírez, Miguel Gualdrón. "To 'Stay Where You Are' as a Decolonial Gesture: Glissant's Philosophy of Antillean Space in the Context of Césaire and Fanon." In *Memory, Migration and (De)Colonisation in the Caribbean and Beyond*, edited by Jack Webb, Rod Westmaas, Maria del Pilar Kaladeen, and William Tantam, 133–152. London: University of London Press; Institute of Latin American Studies, 2020. https://www.jstor.org/stable/j.ctvwh8cwp.14.

Serres, Michel. *Biogea*. Translated by Randolph Burks. Minneapolis: Univocal, 2012.

Serres, Michel. *The Natural Contract*. Translated by Elizabeth MacArthur and William Paulson. Ann Arbor: The University of Michigan Press, 1995.

Serres, Michel with Bruno Latour. *Conversations on Science, Culture, and Time*. Translated by Roxanne Lapidus. Ann Arbor: The University of Michigan Press, 1995.

Sloterdijk, Peter. *Foams: Spheres Volume III: Plural Spherology*. Translated by Wieland Hoban. South Pasadena: Semiotext(e), 2016.

Treanor, Brian. "Mind the Gap: The Challenge of Matter." In *Carnal Hermeneutics*, edited by Brian Treanor and Richard Kearney, 57–73. New York: Fordham University Press, 2015.

Van Keymeulen, Jacques. "Zeeuws Overzee." In *NEHALENNIA* 198 (2017): 18–19. Edited by Pau Heerschap and Rinus Willemsen.

Webster, Jane. "The *Zong* in the Context of the Eighteenth-Century Slave Trade." *Journal of Legal History* 28, no. 3 (November 2007): 285–98. doi:10.1080/01440360701698403.

Wittgenstein, Ludwig. *Philosophical Investigations*. Translated by G. E. M. Anscombe. Oxford: Blackwell, 1953.

Chapter 4

What We Talk About When We Talk About Extinction

Lisa Ottum

Literary works about extinction are difficult to come by. Perhaps it's the abstractness of loss on a global scale or the dearth of familiar figures for vanished species. Or perhaps the trouble lies with form. The lyric mode, for instance, unavoidably signals presence rather than absence—and there is also its inherent anthropocentrism. Other literary forms seem equally unsuitable, tainted by their association with individualism, or heroism—or love. Imagine for a moment a sonnet sequence about extinction: what might such an unlikely expression of desire look like? What emotions would it aim to invoke?

There are, of course, *some* literary works that tackle extinction—including a poem I often teach titled "The Big Picture" (2007) by Ellen Bass. Written in the first-person, the poem opens with a folksy saying: "I try to look at The Big Picture," the speaker begins. She elaborates:

The sun, ardent tongue
licking us like a mother besotted

with her new cub, will wear itself out.
Everything is transitory.[1]

Turning from the distant future to the distant past, she ponders the Cretaceous-Paleogene extinction that wiped out the dinosaurs, and "before that, the volcanoes / of the Permian period" with its innumerable "burnt ferns // and reptiles, sharks, and bony fish" (8–10). *That*, she concedes, "was extinction on a scale / that makes our losses look like a bad day at the slots" (11–12). Still, the speaker cannot "shake [her] longing" for "the last six hundred / Iberian lynx with their tufted ears" and a host of other imperiled species:

Brazilian guitarfish, the 4
percent of them still cruising
the seafloor, eyes staring straight up.

And all the newborn marsupials—
red kangaroos, joeys the size of honeybees—
steelhead trout, river dolphins,

so many species of frogs
breathing through their damp
permeable membranes. (19–27)

On the bus, the speaker "ache[s]" for a glimpse of cardinals in the snow and for "polar bears, the cream and amber / of their fur . . . ." (27, 29–30). Once at home, her thoughts turn to a quotidian scene: sitting on the couch with her teenaged son, watching "A cheap / silver chain shimme[r] across his throat / rising and falling with his pulse" (41–2). "There never was / anything else," she muses, "Only these excruciatingly / insignificant creatures we love" (43–5).

As this sketch of the poem suggests, "The Big Picture" rejects The Big Picture as means of reframing: deep time might contextualize mass extinction, but it cannot account for the speaker's experience. Specifically, it cannot account for affect—for her "longing," for her "ache," and, most importantly, for the love she feels toward her son. Why love one's own kind, or any other kind, in a world destined for destruction? Why linger over desire? And for that matter, why create art? I read the poem's closing lines as a response to such questions. Bass rejects nihilism on the grounds that love makes "The Big Picture" apprehensible in the first place. Extinction is tragic *only* as it relates to love: in the absence of love, species loss is interpretable only against the backdrop of planetary history, a framework in which all losses appear equally meaningless. Either we love species and therefore their persistence means something, or we disavow love in favor of a neutral "big picture." Either there is love or "[t]here never was / anything else." For Bass, love is thus a form of interpretation, an act of reading through which geologic time and the prospect of a sixth mass extinction become meaningful.

Love and affection seldom appear in academic conversations about extinction, where, in general, ugly feelings predominate. Recent work in the environmental humanities examines extinction as it relates to anxiety, mourning, melancholy, and other unpleasant states. Literary critics, for example, question whether traditional genres, such as the elegy, are adequate to the task of confronting a sixth mass extinction.[2] Meanwhile, extinction is increasingly salient to psychologists as a source of psychic distress. Recent studies cite extinction as one of many factors behind "ecological grief," an emerging malady of the Anthropocene.[3] Of course, popular discourse about extinction has long revolved around grief, loss, and scarcity. Since its inception, mainstream environmentalism has framed environmental despoliation of all kinds as a departure from "baseline" conditions.[4] According to this paradigm, anthropogenic changes damage nature's integrity, gradually and irreversibly chipping away at ecosystems. In the early 1980s, Paul and Annie Ehrlich famously imagined Earth as an airplane from which humans are carelessly popping rivets—that is, killing off individual species whose ecosystem function might become clear only in retrospect.[5] This "rivet popper hypothesis" persists even today in the popular imagination, despite evidence that exposes its limitations.[6]

In this essay, I challenge the assumption that only grief, mourning, and melancholy are appropriate responses to the sixth mass extinction. What happens when we reframe extinction in terms of love, affection, and attachment—between and among both humans and nonhuman others? I build on the work of Deborah Bird Rose, Lida Maxwell, and others who have written on love and extinction,[7] as well as the 2009 novel *How the Dead Dream* by Lydia Millet, a text that explores love as it relates to both loneliness and species loss. As Millet's novel illustrates, what we lose with extinction is not only biodiversity but also opportunities for attachment between humans, between humans and animals, *and* between creatures themselves. Extinction is biological impoverishment *and* an attenuation of affect, a dwindling of intra- and interspecies experience. Hence, to linger over love is to insist upon the value of this experience, to resist the zoom out to a disorienting, and ultimately paralyzing "big picture."

The reorientation I propose might seem counterintuitive, or even misguided: why linger over love and affection in the midst of a crisis? Why not instead get down to the cold, hard business of being sad, being mad, and refusing to be placated? As Maxwell observes, love is out of fashion among queer theorists, many of whom have dismissed it as a pernicious force for normativity.[8] Among ecocritics, meanwhile, professing a love of nature is no longer commonplace: if anything, love is associated with the narrative-based, deeply personal scholarship that prevailed during the field's early days. Finally, in Western popular culture, there is perhaps no anti-environmental epithet with more force than "tree-hugger": hugging (i.e., loving) nature is for sentimental hippies, fools who, depending upon your political orientation, are either perverse or simply naïve. Yet what is lost when we marginalize or ignore love in our discussions about extinction? I argue that love has a unique potential to focus attention on the intangible, ineffable losses that accompany extinction. If extinction is a flattening of biodiversity, it is also a flattening of emotional life—a winnowing of opportunities for attachment. E. O. Wilson has characterized the coming era as an Age of Loneliness, or "Eremocene" in which humans will survive the Anthropocene, yet find ourselves alone on a biologically impoverished planet.[9] If we are to avert an Eremocene, then we must foreground the role of love and affection as a better alternative: the opposite of loneliness is companionship, an Age of Camaraderie in which love and a sense of kinship with human and nonhuman others anchors our ethical deliberations.

IT'S OUR GEOLOGICAL EPOCH, AND WE'LL CRY IF WE WANT TO

In 2020, it is difficult, if not impossible, to discuss extinction without invoking the Anthropocene, a concept first articulated in 2000 by chemist Paul Crutzen, and now wielded by humanists, social scientists, and journalists alike. According to the dominant Anthropocene narrative, humans have altered the planet so substantially as to tilt it out of the Holocene, which began about 11,000 years ago, and into a new geological epoch. While disagreement persists among stratigraphers about whether there is, or will soon be, a globally synchronous stratigraphic marker for the Anthropocene, other experts aver that the present extinction crisis is sufficient evidence for an Age of Humans. Indeed, the current "Sixth Extinction" is commonly cited as the key indicator of an Anthropocene: if the planet's past five mass extinction events mark enormous ruptures in Earth's history, then it stands to reason that the ongoing mass extinction of our times will appear similarly catastrophic in retrospect.[10]

By now, a great many thinkers have explored the scientific and philosophical consequences of conceptualizing humankind as a geological force. Persuasive critiques of the Anthropocene abound: as feminist, Marxist, indigenous, and other critics point out, the Anthropocene obscures the role of capitalism, imperialism, and (hetero)sexism in creating our present environmental crises.[11] Less fully explored are cultural manifestations of a "pop Anthropocene," or what literary critic Lynn Keller refers to as the "self-conscious Anthropocene"—that is, our current "period of *changed recognition* when the responsibility humans bear for the condition of the planet and for the fates of Holocene species is widely understood."[12] In Western news publications, one discerns an emerging structure of feeling: environmental changes—including extinction—that were once framed in other terms are now positioned as emblems of the Anthropocene. On one hand, this shift is a positive development: journalists are more explicitly linking habitat loss and other developments to anthropogenic climate change. On the other hand, the mainstreaming of deep time may have dubious effects on how people conceptualize human and nonhuman agency in a time of crisis. The Anthropocene concept rests on the future anterior tense; it is the legacy of humans that will have been many years hence, as seen by observers at some point in the distant future. If the Anthropocene is an era of mass extinction, then the anticipatory memory we are invited to form in the present is one of species already gone: an Age of Loneliness that is already here. Among emotions that one might attach to this "geological" view of life are ambivalence, disorientation, guilt, numbness, and sadness. It is considerably more difficult to attach hope, optimism, or other open-ended emotions to this perspective. Instead, we are forced into the mode of proleptic elegy, mourning for things whose demise feels inevitable.

Arguably, globalized capitalism was already producing an Eremocene long before the term "Anthropocene" emerged by regarding animals—and certain groups of people—as mere resources, rather than companions. Anthropocene discourse also overlooks the richness of our current attachments to species that are (still) here, at least for now, as well as the creaturely attachments and desires unaccounted for by the view from a distant future. What will sea turtles, or sun bears, or other threatened species have wanted in the Anthropocene? This is not a fanciful question: it is, rather, an acknowledgment that the practice of species conservation already brings us into contact with agentic assemblages of people, animals, plants, and natural forces. To frame extinction as a binary choice (to save or not to save) in which the Anthropos watches a finite array of other species-rivets pop one by one off airplane Earth is to ignore the competing inter- and intraspecies interests that inevitably surround conservation. In an article on the so-called Edenic sciences of conservation biology, invasion biology, and restoration ecology, cultural geographers Paul Robbins and Sarah A. Moore take these fields to task for disavowing the political nature of their work. These fields, they argue, face an existential crisis in the Anthropocene, as novel ecosystems and the rapid pace of environmental change challenge the very notion of a "natural" system state. Robbins and Moore therefore urge scientists to "be explicitly, honestly, and strategically political"—to "enunciat[e] desires, while acknowledging those of others."[13] Scientists, they argue, should ask: "What are our desires and how are they entangled in the desires of others? To whom does value flow in [conservation] experiment[s] and at whose expense?"[14] Although affect is not Robbins and Moore's focus, nor are the desires of nonhuman species (as opposed to those of scientists and indigenous peoples), their essay nonetheless chimes with the conceptual experiment I am proposing. Conservation is about desire—ours, and presumably, that of the creatures we hope to save.

My concern is whether the "big picture" afforded by geology allows space for new affective stances toward extinction. As critics of the term "Anthropocene" have demonstrated, the dominant Anthropocene narrative reinforces a pernicious and hetero-patriarchal nature/culture dualism: in this paradigm, "man" acts upon an implicitly feminized nature, thoughtlessly pillaging resources while other species passively await certain doom. Scarcity awaits us over the horizon; our best, and perhaps only option, is to hoard Earth's dwindling number of animals, plants, and insects, and pile them into a lifeboat headed for . . . somewhere. Sarah Ensor contends that the rhetoric of "saving" species evinces our cultural phobia toward "illness, frailty, and death"; in the rush to "save" certain creatures, we have failed to develop an ethic for living with terminality, a challenge queer theory began to explore during the 1980s AIDS crisis.[15] For Ensor, queer texts invite us to ask "what possibilities for relation exist amidst terminality, and what similarities may be between a place (or planet) persisting on 'borrowed time' and a human being (or human couple) doing the same."[16] In addition, they invite us to imagine terminality as its own temporality—as a space of possibility rather than merely a state of "futurelessness." As I explain in the next section, queering the Anthropocene is one way to admit a wider range of emotions into our discussion of extinction and deep time. These include conceptions of ecological love that resonate with non-Western approaches to human and ecosystem health.

WHAT'S LOVE GOT TO DO WITH IT?

Before continuing, it will be helpful to consider the term "love" more closely, for love and related ideas such as desire, empathy, and compassion might not automatically appear compatible with ecological thinking, let alone deconstructing the Anthropocene. According to queer theorist Lee Edelman, the discourse of love abets "reproductive futurism," an ideology in which the figure of the Child is "the perpetual horizon of every acknowledged politics."[17] Under this paradigm, policy choices are directed toward "our children's future"; political discourse "preserve[s] the absolute privilege of heteronormativity by rendering" nonnormative social arrangements "unthinkable."[18] To be sure, mainstream environmentalist discourse invests heavily in reproductive futurism: activists beseech us to take action now in order to save the Earth, or the whales, or the polar bears "for our children." Going green thus affirms normative family structures, and in the process excludes a great many people whose relationship to the future is not biologically reproductive.[19] To brook this impasse, queer ecocriticism has begun to explore alternative models of environmental stewardship that are not bound up in the discourses of procreation or inheritance. Ensor, for instance, proposes a "spinster ecology" that "tend[s] to the future without contributing directly to it."[20] In contrast to the Child, Ensor's spinster "opens for us a window into futurity freed from the bounds of objects, [and] freed from the promise of ever arriving in recognizable form."[21] The spinster adopts "a temporal stance outside the bounds of object-based preservation"; she instead inhabits "the realm of something akin to persistence or continuity."[22] Although Ensor uses the term "care" instead of "love," her concern with attachment challenges the logic that connects love to romance and reproduction, and, in turn, to futurity. Ensor's spinster cares—but in a way that evades the hegemonic continuum linking caring to desire, desire to sex, and sex to children.

Of course, a foundational aim of queer theory is to make room for desires and attachments that fall outside of dominant paradigms—to liberate these concepts from "Western regimes of sexuality—institutions that," in David Hubert's words, "simultaneously enforce heterosexual and intraspecies comportments of desire."[23] In this respect, queer theory intersects with postcolonial theory's critique of the Enlightenment sovereign subject. As Angela Hume and Samia Rahimtoola explain, "[a] key aspect of coloniality was the suppression of indigenous sexual practices deemed to be deviant. Thus, decolonization is closely and necessarily tied to the project of queer liberation."[24] Decolonization, they write, is also about "unsett[ling] normative Western-centric ideas about environmental consciousness."[25] Taken together, these statements suggest that queering the Anthropocene must involve indigenous perspectives on what counts as "love," who can "love," and what "love" accomplishes for both the human and more-than-human worlds. In his recent *Earth Emotions*, environmentalist philosopher Glenn Albrecht develops a typology of "earth emotions" inspired in part by the practices of Australia's Aboriginal people, whose cosmology sees "human life as originating in common ancestors that consisted of human and non-human life in a shared destiny."[26] As Albrecht explains, Aboriginal thought assumes a close connection between people and places, such that one can speak of the country as one's "sweetheart."[27] "Love" in this worldview is not merely a dyadic exchange between people; rather, love encompasses myriad forms of cooperation between and among people, plants, animals, and inanimate matter. In this respect, Western science has only recently caught up with indigenous thinking: Australia's indigenous people have long viewed "the term 'environment'" as a "category mistake," an "erroneous form of dualistic thinking" that arbitrarily separates humans from other beings.[28] Recent science affirms that "bacteria, trees, and humans are not individuals existing as isolated atoms in a sea of competition"; in fact, life itself "consists of commingling microbiomes within larger biomes" such that there are "no clearly defined 'insides' and 'outsides'" to humans, or trees, or anything else. If life is an "arrangement between different types of beings at various scales of existence," then it is dubious to carve "love" into reproductive and nonreproductive forms—into discrete categories such as "affection," "desire," and "care."[29]

Albrecht characterizes loves (*philias*) of various kinds among positive emotions that are "terranascient" or life-affirming, in contrast to destructive, Earth-destroying "terraphthoric" emotions such as ecophobia. We live, he contends, in an era where terraphthoric emotions are increasingly common; in order to address climate change, mass extinction, and other problems, we must embrace the notion of symbiosis itself, a move that involves love for everything from the hydrosphere to the bacteria in our gut microbiomes. The corollary to this assertion is that we must begin to see climate change, mass extinction, and other trends that shatter symbiotic relationships as losses of love. It is this aspect of extinction that literature and other art can illuminate: consider, once again, how Ellen Bass positions love of various kinds as the opposite of species loss. To love something in a decolonializing sense of the term is to affirm its liveness—its embeddedness in dynamic, intersecting symbioses. Seen in the way Albrecht describes, love is not only a feeling, but also a way of naming the forms of "connection, exchange, creation, defense, and sharing" that allow all lifeforms to thrive, from the tiniest microbes to the largest communities of humans (149).

> The principle . . . which animates life in a forest or on the human gut-brain axis, "he writes, is one . . . that is repeated in the love between human beings at a small scale (between lovers) and at the larger scale in the soliphilia or 'love of the whole community' of humans." (149)

In other words, a properly ecological love is not exemplified by the scenes of transcendence depicted in traditional nature writing in which the solitary human subject goes into the wild to discover his unity with something larger than himself. Truly ecological love involves lifeforms larger and smaller than ourselves. It tends toward the promotion of life-creating relationships, and away from forces and systems (such as capitalism and colonialism) that destroy symbioses.

All of this means that a great deal of what passes for "love" in late capitalism is not the love described in "The Big Picture" or other works of art that mourn species loss. Love can take many oppressive forms; it is perhaps a reflexive suspicion of love that has kept it from surfacing in extinction studies. At the same time, the ongoing critical turn away from "paranoid reading" to what Eve Kosofsky Sedgwick refers to as "reparative reading" presents an opportunity to reassess love.[30] What if, instead of reading to expose love as hopelessly repressive, we were to read love reparatively? Susan McHugh proposes something similar in her work on indigenous narratives of platonic cross-species affection. McHugh contends that such stories are essential to the ongoing project of decolonizing literary animal studies, a field which has only recently expanded its efforts beyond Eurowestern and Anglophone texts.

What is more, these stories reveal how "profound ruptures to shared relations between species concern more than losses exclusive to either cultural or biological lifeways."[31] The mass killing of human and nonhuman others is a hallmark of the Anthropocene: Indigenous stories register the coeval violence of genocide and species extinction that has played out for centuries under settler colonialism. At the same time, these stories "validate attachments specific to shared conditions with nonhuman animals, including complex mutual dependencies on other animals and plants unique to particular places."[32] To read love in this context is to recover "lifeways premised on loving relations of cross-species interdependence," worlds that offer the possibility of resistance to the pernicious extractivism that drives mass killing.

Literary representations of ecological love might not engender direct action, although following McHugh, I wager that "stories of multispecies affective bonding" might serve "as an effective means to secure shared futures."[33] The next section offers a brief reading of Lydia Millet's *How the Dead Dream*, a novel that grapples with extinction and other forms of loss as they relate to love. This novel is not "anti-extinction" in any straightforward sense: it is the story of one American man's journey from self-centeredness and an obsession with economic success to his belated realization that he himself is complicit in an unfolding mass extinction. Similar to Bass's poem, the novel's lingering affect is something bittersweet: this is a not a conversation narrative in which love wins the day, but rather a study in love's complexities. This includes the pain associated with loving creatures and forms of symbiosis that might well disappear, or are already disappearing.

BETTER TO HAVE LOVED AND LOST?

How the Dead Dream is not ostensibly "about" extinction, although extinction certainly figures into Millet's vision, as the novel's acknowledgments include a dedication to the West African black rhinoceros ("which disappeared from the world in the time it took to write this book") as well as an alphabetical list honoring "the rarest species in the United States, any of which may vanish in the blink of an eye."[34] *How the Dead Dream* centers on Thomas (called "T."), a white, middle-class American man who comes of age in the 1990s and builds a successful property development company. In his youth, T. is fascinated by money: his first love ("his idol[s]") are the presidents that adorn paper money, whose "venerable" faces and sense of authority fascinate him.[35] Indeed, T.'s attraction to money is the first of several queer attachments—some "terraphthoric" and some "terranascient" to borrow Albrecht's terminology. At six years old, T. feels his allowance money to be "sacrosanct," and in addition to sometimes "secreting coins" in his mouth, he furtively communes with his coin and bill collection at night, his "mind . . . pulsing like a heart" until he feels utterly "spent."[36] As T. grows older, "his love bec[omes] more sophisticated"; rather than physical contact with money, he finds "satisfaction in the surges of energy, in the stream of contact between machines" that process transactions.[37] For T., money not only represents political power, but also the force of life itself: "currency infused all things," he muses, "from the small to the monolithic."[38] T. sees "the noble trace of money" in everything from "the half-imagined bodies of . . . dinosaurs" in the history museum to "the great thick cables that ran beneath the surging Atlantic, [to] the intricate and freezing satellites that whirred a thousand miles above the surface of the earth."[39] The multiplying power of money so dazzles T. that, by the time he is in college, he begins day trading, his mind filled with visions of "a great acceleration."[40]

While T. intuitively grasps capitalism's ecologies, he is less successful reading people, whose inconsistencies and illogical behavior puzzle and annoy him. (T.'s difficulties, combined with his skill at card-counting, might suggest that he is on the autism spectrum). He cannot empathize with his mother's devotion to him, or her religious faith; he is similarly put-off by his peers, regarding social interactions in tactical terms. According to my description thus far, Millet's character might seem a basic antihero who prevails despite his unlikability; however, T.'s evolution resists a standard narrative arc. This is partly because *How the Dead Dream* disrupts the familiar romance plot centered on heterosexual coupling. As a college student, T. perfects his skill at feigning conviviality, while forging no romantic relationships or even close friendships. After graduation, he sails to success in real estate; as a young professional, he "like[s] to be away from people and then suddenly face-to-face," for "all in a rush, they would converge, burning with self-interest like pillars of fire" (33). Only much later in the novel does T. develop anything like empathy, let alone love, for anyone else; importantly, these attachments are brokered by emotional encounters with animals—first, a coyote, then a pet dog, and later endangered zoo animals, toward whom T. begins to feel something like ecological love.

A turning point occurs when T. hits a coyote on the freeway. In shock, he jumps out of his car and runs to the dying animal. As he sits next to the coyote, he is seized by a fit of radical, visceral empathy: he imagines "the shock from the [coyote's] ruined legs coursing through her body, what must be the blind surge of pain as the end closed in." The narrator's description continues:

> A loud end—the rush of cars still distant punctuated by searing noise and glare of those approaching, bearing down viciously and then fading again. She was

> dying in the smells of asphalt, exhaust, and gasoline, no doubt also the smell of her own blood, and him, and other smells he could now know himself.
>
> The fullness, the terrible sympathy![41]

T. wonders if he has ever "felt this [fullness] before."[42] The coyote dies; T. feels "relieved but oddly disoriented. Where was the ambulance? No: he was all she had. All her lights, all her rescue workers." It is "just a coyote," T. reminds himself, "And yet he fe[els] confused" and unable to leave the scene.[43]

This incident introduces several of the novel's themes: animal loneliness, interspecies empathy, and the ethical imperative of witnessing. It is also the beginning of T.'s journey from a hard-nosed capitalist into someone who begins to grasp his own creatureliness. Some weeks after hitting the coyote, he adopts a rescue dog, toward whom he gradually develops a sense of affection. At first, the dog's "superfluous[ness]" confounds him, although he soon looks forward to the end of the workday, when he can return to the dog.[44] His bond with the dog represents the second of his queer attachments, this one tending toward, rather than away from life-affirming: in its very "arbitrar[iness]"[45] his bond with the dog defies conventional expectations, especially alongside his clients' heterosexual conquests. Significantly, T.'s only romantic relationship is abruptly terminated about halfway into the novel, thus foreclosing any gestures toward resolution through reproductive futurism. Not long after adopting the dog, T. falls deeply in love with a woman named Beth who suffers a stroke and dies. Thus, T.'s prospects (and readers' hopes) for a happy ending in which T. settles down and learns to value family over money disappears.

Extinction first comes up about halfway through the novel, when a development project T. is pursuing threatens to wipe out "a small group of kangaroo rats" that are "the last of their kind, on the brink of disappearing."[46] T. works out a mitigation agreement with the local government in which biologists will relocate the rats to another parcel of land; this plan fails when the rats die at a holding facility. Still distraught over the loss of Beth, T. begins spending evenings at the empty mitigation site, and then at zoos, where he searches for solace, and a sense of companionship, among endangered animals. T. imagines these animals' predicament as similar to his own:

> He knew the zoo animals lived in cages but nothing more about them except that they were alone, most of them, not only alone in the cages, often, but alone on the earth, vanishing. Their condition was close to what he was trying to grasp, lay somehow at the base of his growing suspicion that the ground was no longer fixed, was shifting beneath him.[47]

T. realizes that "a loss like his own" is "common" and that he cannot "pretend to the animals' isolation."[48] Still, his feelings for Beth lead him to a kind of interspecies empathy—a "glimpse of something" he finds "in losing Beth": "If a being could be so singular to another, there was no doubt that there was singularity elsewhere, that the irreplaceable nature of being was not limited to his own small circle."[49] It is worth lingering over this moment—because here, the text gestures, if glancingly, toward the value of a queer, anticolonial vision of love, one in which a human relationship is both equated and entangled with other "singular" biotic relationships. T. reasons, and even begins to feel, that if he and Beth enjoyed one "irreplaceable nature of being" as lovers, then similar unique worlds must exist between and among nonhuman creatures.

T. gradually begins to connect extinction with ecological collapse, as well as novel forms of loss and loneliness associated with the diminution of biodiversity. Among these is a "graying" of the world—that is, the transformation of animal habitats as human cities expand. T. flips through magazines where he sees "animals far away," set against "backdrops of green, others yellow, others a bright turquoise. White now and then, Siberia or the Antarctic."[50] In contrast, "there was the gray of human habitation. The blue parts were turning brown, the yellow places to dust, the green places to smoke and ashes."[51] As the gray "metastasize[s]" species blink out, and a "quiet mass disappearance" ensues; T. wonders why this "inversion of the Ark" seems to be "passing unnoticed," except by "fringe elements and elite groups, professors and hippies, [and] small populations of little importance."[52] Here and elsewhere in the text, Millet describes extinction as something rather like the opposite of what Deborah Bird Rose calls "worldmaking," the process by which, "in becoming with others we bring forth worlds of action and meaning, with varied possibilities for life and for death."[53] T. feels as though "Each time one of the animals disappeared—they went by species or sometimes by organizations of species, interconnected—it was as though all mountains were gone, or all lakes. A certain form of the world."[54] As species disappear, possible futures disappear; besides types of life, Earth loses entire ways of being that both include, and exceed, the species lost.

One might argue that T.'s steps toward ecological love come too late—*How the Dead Dream* ends on an unresolved note, with T. lost and delirious in a jungle near his latest development project. In the novel's final pages, T. arrives at "farsighted[ness]."[55] Once in the thrall of capitalism, he realizes that "the market had failed to see the animals for what they were, the animals in their own places with the ancient networks of their culture and landscape intact. Worth far more than single commodities."[56] These "ancient networks" represent "[u]ncountable wealth . . . the kind that was superfluous but the kind that kept you alive, down through the generations."[57] On its own, this revelation might not seem all that profound, especially to readers who are already concerned about the collapse of global ecosystems. More provocative are T.'s ruminations on the "gray" future to come, one in which humans survive the Anthropocene, only to find themselves desperately lonely. This passage of the text is worth quoting at length:

> [W]hen all the ones that you loved had died: so it was, for last animals and for him. What did anyone do, when they were alone? . . . Angry, a person might thrash and fight—even against himself. . . . Later he would have to sit, crying crumple-faced in a pile of the dead. This was how it would be with men, after they finished their work. They would have to be alone after that, and for so many years. On and on they would live, surrounded by gray. Complexity would be gone, replaced with dull sameness that stretched out unending . . . and when they had killed all of their friends and everywhere was empty: only then would they see how terribly they had loved them.[58]

In T.'s vision, humans fail to acknowledge their love for other creatures until after we have already killed them off. We discover in retrospect that the tragedy of extinction is not merely about biodiversity—it is also about love and companionship, about infinite modes of interspecies worldmaking apprehensible only through affect. Implied is a corollary: if we attuned to our love for nonhuman others now, we might avert a mass extinction.

In the final pages of *How the Dead Dream*, as T. drifts in and out of sleep, an animal he does not recognize—something with "tough skin" and "coarse hair"—lies down on top of him and sleeps.[59] Wondering what the animal "could . . . want from him," T. imagines "how it must have felt once—an imprint of the touch of other bodies, how they wriggled close there, in the den." The animal "breathe[s] thickly," and "for a time," T. does know what the animal feels.[60] "They both [have] lungs" and "love to sleep"; they "lik[e] to be alongside each other in the comfort of their rhythm."[61] T. concludes that the animal has mistaken him for its mother—that having found companionship, the animal is "glad" and at peace. "When it comes to the future," T. thinks, "the animal might not have plans," though it has impressions, "the impression of seasons . . . the feel of others like itself here close, of others like itself further out. . . ."[62] This is perhaps the queerest, and also the most "ecological" moment in the book. Lying on the jungle floor, T. finds his own sense of bodily boundaries dissolving, whatever remaining love he feels now attached to an unknowable, and yet companionable animal other. He is united with this animal by the feeling that "home [is] flesh, [is] nearness."[63] This is not a moment of human spiritual transcendence through nature, but rather a humbling, provisional experience of shared vulnerability. It gestures, for one fleeting second, to the kind of love we might need to combat extinction: a love that seeks cooperation and endurance in the midst of precarity.

SOME CONCLUSIONS

I am not so naïve—nor, I imagine, is Millet—to think that averting the looming extinction crisis could be as simple as asking people to love the world's creatures. Love is not a policy solution, and in general, the discourse of love has a mixed record when it comes to political movements. Still, it is worth asking how love can help us to navigate the moment we inhabit. Instead of asking "What will we lose in the future?" what if we were to ask: "What will we love in the future?" What kinds of flourishing multispecies environments will best support the emotional lives we desire? In addition, we might ask how interspecies assemblages already support and enrich our loving human-to-human relationships. Doing so will bring into relief all that we stand to lose if extinction continues at its current rate—all the many feelings and experiences unaccounted for by endangered species lists, by aerial photos of the Amazon, and by placards at the zoo.

Timothy Morton notes that although "[r]easons for being nice to other life-forms abound" there is "around them . . . a ghostly penumbra of feelings of appreciating them for no reason at all."[64] He continues: "[j]ust loving something never has a great reason attached to it. If you can list all of the reasons why you 'should' love this particular person, you are probably not in love."[65] Perhaps, when it comes to extinction, there is something to be said for love's stubborn unjustifiability. Love does not need scientific sanction; it does not require ratification by experts. One can love an animal, or a plant, or an insect without recourse to its ecological function, or some other external measure of value. In this moment of environmental crisis, love is clearly not all we need—still, it is a vital addition to our toolkit, for it can help us to envision the possible futures surrounding extinction in terms of abundance and desire, rather than (only) scarcity, fear, and anxiety. The big picture thrills our brains, but it is the wrong size for our hearts.

NOTES

1. Ellen Bass, "The Big Picture," in *The Human Line* (Copper Canyon Press, 2007), 74 (lines 1–5). Subsequent references to the poem are cited in-text by line number.

2. For examples of this discourse, see Ursula Heise, *Imagining Extinction: The Cultural Meanings of Endangered Species* (Chicago: University of Chicago Press, 2016), 19–54; Margaret Ronda, *Remainders: American Poetry at Nature's End* (Stanford: Stanford University Press), 91–112; and Jesse Oak Taylor, "Mourning Species: *In Memoriam* in an Age of Extinction," in *Ecological Form: System and Aesthetics in the Age of Empire*, ed. Nathan K. Hensley and Phillip Steer (New York: Fordham University Press, 2019), 42–63.

3. See Ashlee Cunsolo and Neville R. Ellis, "Ecological Grief as a Mental Health Response to Climate Change-Related Loss," *Nature Climate Change* 8, no. 4 (2018), 275–281, DOI 10.1038/s41558-018-0092-2.

4. See Lesley Head, *Hope and Grief in the Anthropocene: Re-conceptualizing Human-Nature Relations* (New York: Routledge, 2016), 38–53.

5. Paul Ehrlich and Anne Ehrlich, *Extinction: The Causes and Consequences of the Disappearance of Species* (New York: Random House, 1981), xi.

6. The concept of "novel ecosystems" complicates the rivet popper hypothesis. For more on the Ehrlich's airplane metaphor and novel ecosystems, see Eben Kirksey, *Emergent Ecologies* (Durham: Duke University Press, 2016), 2–4; 214–218.

7. See Deborah Bird Rose, *Wild Dog Dreaming: Love and Extinction* (Charlottesville: University of Virginia Press, 2011); Lida Maxwell, "Queer/Love/Bird Extinction: Rachel Carson's *Silent Spring* as a Work of Love," *Political Theory* 45, no. 5 (2017), 682–704, DOI 10.1177/0090591717712024; Tom van Dooren, *Flight Ways: Life and Loss at the Edge of Extinction* (New York: Columbia UP, 2014); and Heise.

8. Maxwell, "Queer/Love/Bird," 686–687; see also Lydia Maxwell, "Does Love Have a Politics?" in *The Los Angeles Review of Books*, February 4, 2016, https://lareviewofbooks.org/article/does-love-have-a-politics/.

9. See Edward O. Wilson, *Half Earth: Our Planet's Fight for Life* (New York: Liveright, 2016), 20.

10. For an accessible overview of the history of extinction science, see Elizabeth Kolbert, *The Sixth Extinction: An Unnatural History* (New York: Picador, 2014).

11. Other common critiques of the Anthropocene are that it naturalizes destructive environmental choices by aligning them with "human" nature; that it obscures from humans' creatureliness; and, perhaps most troublingly, that it elides differentials in vulnerability to pollution, sea level rise, food insecurity, and other unfolding disasters. There is not space in this essay to offer a comprehensive overview of responses to the dominant Anthropocene narrative; for a trenchant critique of the Anthropocene, see Christophe Bonneuil and Jean-Baptiste Fressoz, *The Shock of the Anthropocene: The Earth, History, and Us*, trans. David Fernbach (New York: Verso, 2016); see also Stefania Barca, *Forces of Reproduction: Notes for a Counter-Hegemonic Anthropocene* (Cambridge: Cambridge University Press, 2020). doi: 10.1017/9781108878371 and Richard Grusin, ed., *Anthropocene Feminism* (Minneapolis: University of Minnesota Press, 2017).

12. Lynn Keller, *Recomposing Ecopoetics: North American Poetry of the Self-Conscious Anthropocene* (Charlottesville: University of Virginia Press, 2017), 2.

13. Paul Robbins and Sarah A. Moore, "Ecological Anxiety Disorder: Diagnosing the Politics of the Anthropocene," *Cultural Geographies* 20, no. 1 (2012), 14.

14. Ibid., 13.

15. Sarah Ensor, "Terminal Regions: Queer Ecocriticism at the End," in *Against Life*, ed. Alastair Hunt and Stephanie Youngblood (Evanston: Northwestern University Press, 2016), 42.

16. Ibid., 44.

17. Lee Edelman, *No Future: Queer Theory and the Death Drive* (Durham: Duke University Press, 200–4), 3.

18. Ibid., 2.

19. As Noël Sturgeon notes, hegemonic forms of the "normative" family are also "bound up in environmentally dangerous social and economic practices" (107). Sturgeon invites environmentalists and ecocritics to embrace a more expansive conception of "reproduction" that not only includes biological reproduction, but also forms of social and economic (re)production in order to discover "the socioeconomic and sociopolitical arrangements best suited for successful and sustainable reproduction on the biological, social, and environmental levels." See Sturgeon, "Penguin Family Values: The Nature of Planetary Environmental Reproductive Justice," in *Queer Ecologies: Sex, Nature, Politics, Desire*, ed. Catriona Mortimer-Sandilands and Bruce Erickson (Bloomington: Indiana University Press, 2010), 108.

20. Sarah Ensor, "Spinster Ecology: Rachel Carson, Sarah Orne Jewett, and Nonreproductive Futurity," *American Literature* 84, no. 2 (2012), 409.

21. Ibid., 419.

22. Ibid., 429.

23. David Hubert, "The Equine Erotopoetics of Linda Hogan and Joy Harjo," *Interdisciplinary Studies in Literature and the Environment* 25, no. 1 (2018), 169.

24. Angela Humen and Samia Rahimtoola. "Introduction: Queering Ecopoetics," *Interdisciplinary Studies in Literature and the Environment* 25, no. 1 (2018), 143.

25. Ibid., 142.

26. Glenn Albrecht, *Earth Emotions: New Words for a New World* (Ithaca: Cornell University Press, 2019), 6.

27. Ibid., 145.

28. Ibid., 101.

29. Ibid., 116.

30. See Eve Kosofsky Sedgwick, *Touching Feeling: Affect, Pedagogy, Performativity* (Durham: Duke University Press, 2003), 123–152.

31. Susan McHugh, *Love in the Time of Slaughters: Human-Animal Stories Against Genocide and Extinction* (University Park: Penn State University Press, 2019), 2.

32. Ibid., 16.

33. Ibid., 9.

34. Lydia Millet, *How the Dead Dream* (New York: Mariner, 2009), 245.

35. Ibid., 1–2.

36. Ibid., 8.

37. Ibid., 13.

38. Ibid., 14.

39. Ibid., 14.

40. Ibid., 18.

41. Ibid., 36.

42. Ibid., 36.

43. Ibid., 38.

44. Ibid., 40.

45. Ibid., 40.

46. Ibid., 123.

47. Ibid., 134.
48. Ibid., 166.
49. Ibid., 166.
50. Ibid., 139.
51. Ibid., 139.
52. Ibid., 139.
53. Rose, *Wild Dog Dreaming*, 11–12.
54. Millet, *How the Dead Dream*, 139.
55. Ibid., 238.
56. Ibid., 238.
57. Ibid., 238.
58. Ibid., 242.
59. Ibid., 243.
60. Ibid., 243.
61. Ibid., 243.
62. Ibid., 243.
63. Ibid., 244.
64. Timothy Morton, *Being Ecological* (Cambridge: MIT Press, 2018), 126.
65. Ibid., 126.

BIBLIOGRAPHY

Albrecht, Glenn. *Earth Emotions: New Words for a New World*. Ithaca: Cornell University Press, 2019.

Barca, Stefania. *Forces of Reproduction: Notes for a Counter-Hegemonic Anthropocene. Elements in Environmental Humanities*. Cambridge: Cambridge University Press, 2020. doi:10.1017/9781108878371.

Bass, Ellen. “The Big Picture.” In *The Human Line*, 74–78. Port Townsend: Copper Canyon Press, 2007.

Bonneuil, Christophe and Jean-Baptiste Fressoz. *The Shock of the Anthropocene: The Earth, History, and Us*. Translated by David Fernbach. New York: Verso, 2016.

Cunsolo, Ashlee and Neville R. Ellis. “Ecological Grief as a Mental Health Response to Climate Change-Related Loss.” *Nature Climate Change* 8, no. 4 (2018), 275–281. doi: 10.1038/s41558-018-0092-2.

Edelman, Lee. *No Future: Queer Theory and the Death Drive*. Durham: Duke University Press, 2014.

Ehrlich, Paul and Anne Ehrlich. *Extinction: The Causes and Consequences of the Disappearance of Species*. New York: Random House, 1981.

Ensor, Sarah. “Spinster Ecology: Rachel Carson, Sarah Orne Jewett, and Nonreproductive Futurity.” *American Literature* 84, no. 2 (2012), 409–435.

Ensor, Sarah. “Terminal Regions: Queer Ecocriticism at the End.” In *Against Life*, edited by Alastair Hunt and Stephanie Youngblood, 47–61. Evanston: Northwestern University Press, 2016.

Grusin, Richard, ed. *Anthropocene Feminism.* Minneapolis: University of Minnesota Press, 2017.

Head, Lesley. *Hope and Grief in the Anthropocene: Re-conceptualizing Human-Nature Relations*. New York: Routledge, 2016.

Heise, Ursula. *Imagining Extinction: The Cultural Meanings of Endangered Species.* Chicago: University of Chicago Press, 2016.

Hubert, David. "The Equine Erotopoetics of Linda Hogan and Joy Harjo." *Interdisciplinary Studies in Literature and the Environment* 25, no. 1 (2018), 169–185.

Hume, Angela and Samia Rahimtoola. "Introduction: Queering Ecopoetics." *Interdisciplinary Studies in Literature and the Environment* 25, no. 1 (2018), 134–149.

Keller, Lynn. *Recomposing Ecopoetics: North American Poetry of the Self-Conscious Anthropocene*. Charlottesville: University of Virginia Press, 2017.

Kirksey, Eben. *Emergent Ecologies*. Durham: Duke University Press, 2015.

Kolbert, Elizabeth. *The Sixth Extinction: An Unnatural History*. New York: Picador, 2014.

Maxwell, Lydia. "Does Love Have a Politics?" In *The Los Angeles Review of Books*, February 4, 2016, https://lareviewofbooks.org/article/does-love-have-a-politics/.

Maxwell, Lydia. "Queer/Love/Bird Extinction: Rachel Carson's *Silent Spring* as a Work of Love." *Political Theory* 45, no. 5 (2017), 682–704. doi: 10.1177/0090591717712024.

McHugh, Susan. *Love in the Time of Slaughters: Human-Animal Stories Against Extinction*. University Park: Penn State University Press, 2019.

Millet, Lydia. *How the Dead Dream: A Novel*. New York: Mariner, 2009.

Morton, Timothy. *Being Ecological*. Cambridge: MIT Press, 2018.

Robbins, Paul and Sarah A. Moore. "Ecological Anxiety Disorder: Diagnosing the Politics of the Anthropocene." *Cultural Geographies* 20, no. 1 (2012), 3–19. doi: 10.1177/1474474012469887.

Ronda, Margaret. *Remainders: American Poetry at Nature's End*. Stanford: Stanford University Press, 2018.

Rose, Deborah Bird. *Wild Dog Dreaming: Love and Extinction*. Charlottesville: University of Virginia Press, 2011.

Sedgwick, Eve Kosofsky. *Touching Feeling: Affect, Pedagogy, Performativity*. Durham: Duke University Press, 2003.

Sturgeon, Noël. "Penguin Family Values: The Nature of Planetary Environmental Reproductive Justice." In *Queer Ecologies: Sex, Nature, Politics, Desire*, edited by Catriona Mortimer-Sandilands and Bruce Erickson, 100–133. Bloomington: Indiana University Press, 2010.

Taylor, Jesse Oak. "Mourning Species: *In Memoriam* in an Age of Extinction." In *Ecological Form: System and Aesthetics in the Age of Empire*, edited by Nathan K. Hensley and Phillip Steer, 42–63. New York: Fordham University Press, 2019.

van Dooren, Thom. *Flight Ways: Life and Loss at the Edge of Extinction*. New York: Columbia University Press, 2014.

Wilson, Edward O. *Half Earth: Our Planet's Fight for Life*. New York: Liveright, 2016.

Chapter 5

Rat-Fall

Time and Taxa in the Colorado River Delta, c. 1900

Alex Benson

> "A most excellent 'rat-fall' may be made of a strong barrel, about half full of water. The cover should be placed on a pivot and well baited."
> —Yearbook of the United States Department of Agriculture, 1896

On June 30, 1895, Dane Coolidge writes a letter home. He is with a zoological expedition from Stanford University, camped near Silver Lake, California, in the Sierra Nevada. His letter describes the routine of the trappers' days, their concerns about supplies, their next destination. Mid-correspondence, though, he pauses, lifting his eyes. "I see a little chipmunk on the fence now," he writes, "and will stuff his skin to-morrow; we don't work Sundays."[1] Skipping from the sight of the specimen "now" to its taxidermic stuffing at the beginning of the workweek, the sentence elides the process of trapping, killing, and skinning.[2] It's a syntactic analog of the click with which photography has been said to shuttle its subject from life to death (figure 5.1).[3] The animal may be on the fence (a symbol of indeterminacy, an index of territorial enclosure), but its fate is so certain it doesn't need to be written.

Perhaps, though, this ascribes too much predictive power to the naturalist. Say he does catch a chipmunk the next day. How will he know that it is the same one, and that this one didn't get away? Practically speaking, of course, it hardly matters. The sardonic, macho confidence of the claim is founded on an imprecise but potent synecdoche: if the very reason to trap a given specimen is its representation of a larger whole, and if therefore there exist other parts which are similarly representative of the type, and if these parts have an especially robust fungibility when you're getting paid by the mammal, then

Figure 5.1 Ground Squirrel Photographed by Dane Coolidge. *Source:* "White-tailed spermophile (*Spermophilus leurcurus* [*sic*])," in Witmer Stone and William Everett Cram, *American Animals: A Popular Guide to the Mammals of North America North of Mexico, with Intimate Biographies of the More Familiar Species* (New York: Doubleday Page and Co., 1902), facing p. 162.

skinning one is something like skinning them all. In this way, the doom of this chipmunk—a doom inseparable from its typicality as Chipmunk—starts to take on the uncanny rhetorical shape of a little extinction.[4]

This chapter is about that shape, about the co-constitutive claims of taxonomy and temporality that give narratives of endangerment and extinction their legibility, their force, and their portability across lines of human and other-than-human life. Such claims are not politically neutral. Scholarship in animal studies has long registered the ideological complexities of wildlife conservation discourse and other sites of biological valuation. Critical work in Indigenous studies has for even longer elaborated the ways that the anthropological imagination of settler colonialism selectively designates living populations as "future ghosts," producing "a landscape of perpetual vanishing."[5] The text around which my argument here revolves, a work of fiction titled "The Biologist's Quest" by John M. Oskison, finds the overlap of these spaces of critique in the story of a specimen collector and his guides; it finds a landscape of both vanishing and perpetuation in the Colorado River borderlands; and it finds ghosts and futures in the figure of a short-tailed rat. My attention will move from the details of this text to their confluence with conventions of ethnographic writing, with the zoology of North American rice rats, and with narratives of land and labor that are inseparable from these scientific histories.

The chipmunk on the fence is a bit player in the story behind this story. Oskison (1874–1947), a citizen of the Cherokee Nation who grew up on a ranch outside Vinita in Indian Territory (now Oklahoma), met Coolidge (1873–1940), the white son of a southern California orange farmer, while studying law at Stanford. They both edited and wrote for the university literary magazine, the *Sequoia*. After graduating in 1898, they both studied writing for a year at Harvard. In 1900, they traveled together in western Europe, with Oskison later recalling that, during the transatlantic passage, he listened while Coolidge told him story after story about his life.[6] To pay for the trip, Oskison had a "pile" of over 400 dollars saved up; Coolidge met his costs by catching over 400 rodents for sale to the Smithsonian.[7] If this was an eccentric approach to the grand tour, it was the same way Coolidge had defrayed the costs of his English degree. During term breaks, he trapped for various natural history institutions, including the university. He took fieldnotes in a notebook meant for lectures, with "Mammals" handwritten on the cover.[8] On one trip, when a stage driver delivered the wrong grade of cotton, he griped that "the driver should take Hudson's course on 'interpretation'"—referring to William Henry Hudson, who specialized in Romantic poetry and aesthetic theory.[9] On occasion, Coolidge's notes themselves take an aesthetic turn:

> Below beautiful, sharp ridges covered with different oaks and madronas ran down to sand wash, broad, dry, shiny, cactus + brush on sides, no house no fence, one trail down wash, the sand-bed meandered, like the course of ones life.[10]

Published in the July 1901 issue of San Francisco's *Overland Monthly* magazine, "The Biologist's Quest" plays on the experiences of Oskison's naturalist classmates, especially but not only Coolidge. The story begins with a zoological hunch: some scientists at the Smithsonian believe that "a certain species of short tailed rat," previously designated extinct, might still inhabit the delta of the Colorado River, where it flows from the US-Mexico border and through Quecha (Yuma) and Cucapá lands to the Gulf of California.[11] A white naturalist named Lake is commissioned to investigate, and in Yuma, Arizona, he hires an "old Yuma Indian" named Kitti Quist and a "Mexican desert guide," Joe Maria, to take him down the river (52).[12] Once they have reached the gulf, Lake debarks and walks alone to the dunes where he hopes to find the rat. He never does. When weather forces the guides to bring the boat farther south than their planned meeting point, Lake wanders, becomes severely dehydrated, and, nearing death, dreams of the "wonderful short-tailed rat, swimming forever from bank to bank of a sluggish salt pool that rose and fell as the tide crept in and out" (55). The guides rescue him, and on the way back to Yuma, Lake announces that he is done forever with "the short tailed rats" (57). This renunciation may reflect a sentiment Coolidge expressed in France in 1900; "I am going to retire from this rat business," he wrote to his father from Barcelonette.[13] But his influence on the text gets a more direct nod in a closing mention of the similarly named zoologist who ends up buying Lake's gear: "Cooley . . . went down to Yuma . . . to catch chipmunks for the new zoological park in New York" (57). In 1899, Coolidge followed the same itinerary, "Yuma and southward" and trapped "live mammals and birds" for Central Park.[14]

Transforming his source material, Oskison frames his fictional version of such an expedition within an understated farce of zoological debate. If Lake finds a specimen of the short-tailed rat in question, disproving the extinction claims, it will be "a curious survival, and the scientist who could secure and classify it would earn an enviable reputation" (52). Lazarus taxon! If he does not, presumably the case for this species' extinction would be supported. But the evidentiary burden is asymmetrical. Survival claims can point to forms of positive physical demonstration (among them, paradoxically, taxidermized corpses) that extinction claims, tasked with proving an absence, don't as readily afford. So although Lake fails to find the rat, the debate should, one would think, remain open. Biologists may have developed statistical methods of generating fairly reliable extinction claims, but a single, understaffed, abbreviated, hallucinatory excursion through unfamiliar terrain would not strike most observers as conclusive, a century ago or today. However, in the end, the case is abruptly closed, the epistemological terrain shifting underfoot as the debate is reframed in terms of another question entirely, not *does it still* but *did it ever:* "Professor McLean, of the Pennsylvania Scientific Society, published a pamphlet in the fall of 1897 to show that the short-tailed rat described by the Smithsonian never existed except in the imagination" (57). The dismissal of "imagination" presents the claim as the authoritative correction of a fiction, but McLean's "pamphlet" sounds a little lean, a jealous play in the short-tailed-rat reputation game.

The narration does not specify the substance of McLean's argument, but the likely inference is that the scholar has subsumed this rat under another species designation, having judged it appropriate to collapse some phylogenetic distinction.[15] Such a claim shifts the question of the species' life (or, more precisely, whatever set of lives had been imagined to constitute this species) from when to what, from the diachronic to the synchronic. The shift is also from a question that threatens to remain open indefinitely to one that admits of categorical decision. This produces nested conclusions: the closing of the debate at the end of the story.

If it's true of the experience of narrative structure that "the very possibility of meaning plotted through sequence and through time depends on the anticipated structuring force of the ending," then this particular ending produces an odd effect.[16] On one hand, it consolidates Oskison's satire of natural history, in that the entire expedition is made pointless. (Even if Lake had found the rat, it wouldn't have interested anyone, its life having been rendered unremarkable as just another example of some other species designation that lacks the frisson of possible disappearance.) On the other hand, in suggesting, as part of the same satire, that perhaps the scientific debate should not have been so summarily concluded—that "enviable reputation" may have biased it—the story's ending provokes us to think twice about story-endings as well. Implying that the "structuring force" of a conclusion, the sense of definitional synchrony that comes with the closing of a case, may involve a situated expression of power or vanity, Oskison's ending opens a horizon for other, less teleological, temporalities.

In fact, though, the text has already opened that horizon in multiple ways. It does so, first of all, through its emplotment. The story first follows the biologist's itinerary, beginning with his arrival in Yuma, in a relatively linear fashion, though with some moments of retrospection to indicate characters' backgrounds. But once Lake dreams of the rat and collapses on the verge of death, the text breaks from this chronological mode, interrupting itself with a line of asterisks and backtracking to explain what the guides have been up to during the same interval, before finally uniting these story threads at the moment of rescue. And then the end brings us back to the beginning: Lake, who had replaced a prior naturalist, will now be replaced by Cooley. (Will they ever stop coming?) While this may not in itself sound like such a radical narrative structure, there's a lot at stake, for Oskison, in such spiralizing movements, in the asynchronies of this text as an act of storying.[17] And we can begin to get a sense of these consequences—of how these temporalities mediate relations of human and nonhuman mortality—by attending to passages of the text that, by deferring the central question of a surprising survival, dwell on the fatal procedures of taxidermy and of a peculiar version of snake dancing. So I will consider each of those moments before returning to the short-tailed rat and to the imaginative genealogies of its classification and its threat status.

GENTLE ART, ROUGH MAGIC

> "I notice you use the term 'rat-fall' in your précis; would one dead rat come under that category?—There may have been one dead rat in a house, but two or three is more general."
>
> —Transcripts of the Plague Commission at Belgaum, India, 1899

Camped by the Colorado one evening, Lake shows his Yuma guide how he performs taxidermy, catching and skinning some lizards and a desert rat (length of tail unspecified). Kitti Quist watches "with astonishment the transformation from a limp corpse to a flabby, empty skin, then back again to a cotton-stuffed, perfectly shaped rat, pinned out in a scampering attitude" (53). The paradoxes here of a body being at the same time "pinned" and "scampering," and of death as a means to renewal, have been contextualized by cultural studies of the taxidermic imagination in the early-twentieth-century United States. If (per Donna Haraway) taxidermy's production of an eternal Edenic present—vitality in stasis—is meant to alleviate anxieties of social decadence, appealing particularly to racial eugenicists and patriarchists, it also (per Fatimah Tobing Rony) tropes the ethnographic desire to make that which is dead seem living, in the way that salvage anthropology neglects contemporaneous social processes in favor of more "authentic" forms of traditional practice understood to be endangered.[18]

The association of naturalist practices with ethnography did not go unrecognized in Oskison's circles. Commonplaces related to fieldwork and to the observation of habitual behavior bridged natural-scientific and social-scientific discourse. An 1899 article about Coolidge, for instance, noted that his collection of specimens "requires that he visit many different localities," where "he must remain long enough to become thoroughly acquainted with the habits and the life history of his victims."[19] Stanford did not yet have a dedicated anthropology department when Oskison and Coolidge studied there, but in the *Sequoia*, edited by students, one could find discussions of ethnology alongside folklore, zoology, travel writing, and the kinds of local color sketches that were highly popular in the period and that make up the bulk of Oskison's early work.[20] In a March 1895 piece for the magazine, Mabel L. Miller paraphrased Franz Boas' suggestion, in a recent lecture at Stanford, "that an interesting history was yet to be written of the almost extinct tribes of the Pacific Coast."[21] Describing her study of the "burial and mourning customs" of an unspecified people that "will doubtless soon become extinct" (288), Miller equivocates: the piece describes the group as "*once* inhabiting the east side of the Sacramento River" while also indicating that they still do, so that when Miller writes that "they *had* many customs of mourning" (289, my emphases) the past tense comes to imply the loss of either object or subject, customs or people. This equivocation collapses into redundancy. The dead don't mourn.

The turn of the twentieth century was a moment of particular intensity for such discourse about vanishing Indians, a "dominant discourse," as Wesley Leonard notes in this volume, replicated materially in the settler state's facilitation of programs of linguistic assimilation, genocidal violence, and dispossession. These programs included the parceling and often the expropriation of communally held lands by means of territorial allotment. In Indian Territory, Oskison had a close view of this process as carried out under the protocols of the Curtis and Dawes Act. These territorial transformations figure throughout his writing. His political orientation to such histories, though, has eluded scholarly consensus. Does he, in the terms with which Daniel Heath Justice coordinates the politics of Cherokee literary history, entirely follow a Beloved path of peace and compromise, or does his work also encompass Chickamauga principles of active resistance?[22] Or is he simply assimilationist? A major factor in scholars' collective difficulty with this question stems from Oskison's strong inclinations, as a writer, toward ironic expression, and the ironies of "The Biologist's Quest" hardly confirm any single view of its author's politics. They do, though, add up to a pointed reflection on the medium of fiction as a site for both cultural representation and interspecies ethics.

Take one of the story's central puns: short tail, short tale. Although it is a rote bit of wordplay, here it interfaces complexly with genre and geography. The pun frames the zoological folly over the potentially extinct animal as a joke about the medium of the short story, the privileged venue for local color fiction and its representations of folks whose ways of life are imagined to be imminently plowed under by an industrialized, mass-mediated modernity. The scene evoked: urban magazine writers heading down the river, clueless, in search of short tales.

If those hypothetical writers were to actually mount such an expedition, Kitti Quist might be there to guide them, for a fee. In this character, Oskison explores forms of survival and invention in and against the currents of westward settler incursion, while opening a satiric view onto associated scripts of cultural performance and labor. Consider the scene in which the guide explains his various past employments (each evocative vignette, in itself, the shortest of tales). Although Kitti Quist had once "been the most feared medicine man in the Southwest," this changed as "the Yumas grew poorer, less energetic, and careless of the fame of their great man." To get by, Kitti Quist performs for tourists; he serves as a guide for miners, smiling at their failure; then, after curing the governor of rheumatism by sucking his joints, he becomes a "self-important white man's medicine doctor." And "now he was going to help the new doctor catch rats—for what he knew not. And next he would be?—well, he didn't know" (52–53). Yet his itinerancy is not accompanied by dread. When Lake is driven to the point of insanity, Kitti Quist takes the naturalist's tools and sells them to the next comer. He may not know what he will be (at least, he isn't saying), but—in a casual assumption of personal futurity—he will be something. His name itself seems to announce this adaptability in its lexical resonances: the first part sneaks a feline homonym into a story about chasing a rodent, while the second part stands in close proximity to the title's "quest." In fact, the original printing of the story includes a single instance of Kitti Quist with an alternate spelling that underscores this proximity: "'You have showed me strange medicine tricks with the rat and the lizards,' said Kitti Quest" (53).[23]

Reconsidered with this echo in mind, the title "The Biologist's Quest" takes on a secondary sense. There's a familiar anthropological grammar of possession here: my tribe, my informant. And indeed, after Lake shows Kitti Quist (I'll retain the predominant spelling) his "strange medicine tricks," the guide's response might seem to follow the scripts of salvage. Astonished by the scientific transformation of the rat, he in turn finds a rattler and demonstrates a snake dance.[24] He describes this as a tradition in danger of disappearance, boasting that his version is still robust whereas other (Yaqui) dancers in the region have resorted to performing a watered-down version with young snakes (53). But he presents an idiosyncratic version of snake dancing—one that, I'd suggest, Oskison imagines for particular rhetorical purposes rather than for the verisimilar representation of southwestern dance practices—and the process of the performance suggests a form of what Audra Simpson calls "ethnographic refusal"; this dance, in other words, is an act of analysis rather than of acquiescence.[25]

When Kitti Quist first grabs the snake by the neck, it writhes and hisses. He calms it with slow movements until "all motion had ceased; the rattler lay along his hand and arm pliant and quiet as a huge cord; the unwinking eyes were still and the rattling had ceased" (53). ("To make the rattlesnake pose—that is the heavy work of the artist," wrote Coolidge in a 1908 article titled "The Gentle Art of Photographing Rattlesnakes.")[26] Eventually, after untwining the snake, Kitti Quist recomposes himself, then casually takes the snake by the tail. "He whirled it about his head and brought it back with a jerk that separated the head and body, and flung the mutilated trunk away" (53). In a sequential inversion of taxidermy, here the living is first rendered silent, motionless, as if dead. Only then is it killed. This sequence figures the relation between myths of vanishing and material violence. To be seen as already dead is to be exposed to injury without redress. But the mutilation of the snake also presents a more straightforward contrast with taxidermy. No longer any good for reconstruction, study, cataloging, or display, its body becomes unavailable to the specimen economy.[27]

The same is true of a rat found only in a dream. After the biologist and the guides have reached the gulf, Lake walks into the hills to search; meanwhile, weather forces Kitti Quist and Joe Maria to steer the boat a ways off the coast. Lake is left stranded overnight, runs out of food, gets dehydrated, and finds himself walking miles farther than expected because of the unique topography of the delta—those "desolate, saline mud flats, ten to twenty miles wide and forty or fifty miles long, intersected by meandering sloughs" (meandering like the course of one's life?) through which, as it happens, Coolidge's colleague William Wightman Price had stumbled in 1898 (figure 5.2). According to a later account, having traveled down the Colorado and "penetrated into the gulf with some Mexicans in a ramshackle craft," Price grew so seasick that he tried to walk back to Yuma alone, almost dying of dehydration before he "reached an Indian settlement."[28] Lake's itinerary largely mirrors Price's, but (in addition to the difference of the rescue) no accounts of Price's journey include a short-tailed rat, nor a dream of it swimming in a tidal slough; these elements are specific to Oskison's text.[29] The effect of their inclusion is to make the failure of the quest issued by the Smithsonian into an unintentional fulfillment of the kind of "vision quest" practiced by Indigenous peoples in the Southwest and other parts of North America—given, that is, the possibility of such fulfillment without intention.[30] Oskison is again playing fast and loose with such cultural reference, in a parodic mode that is visible in the mismatch

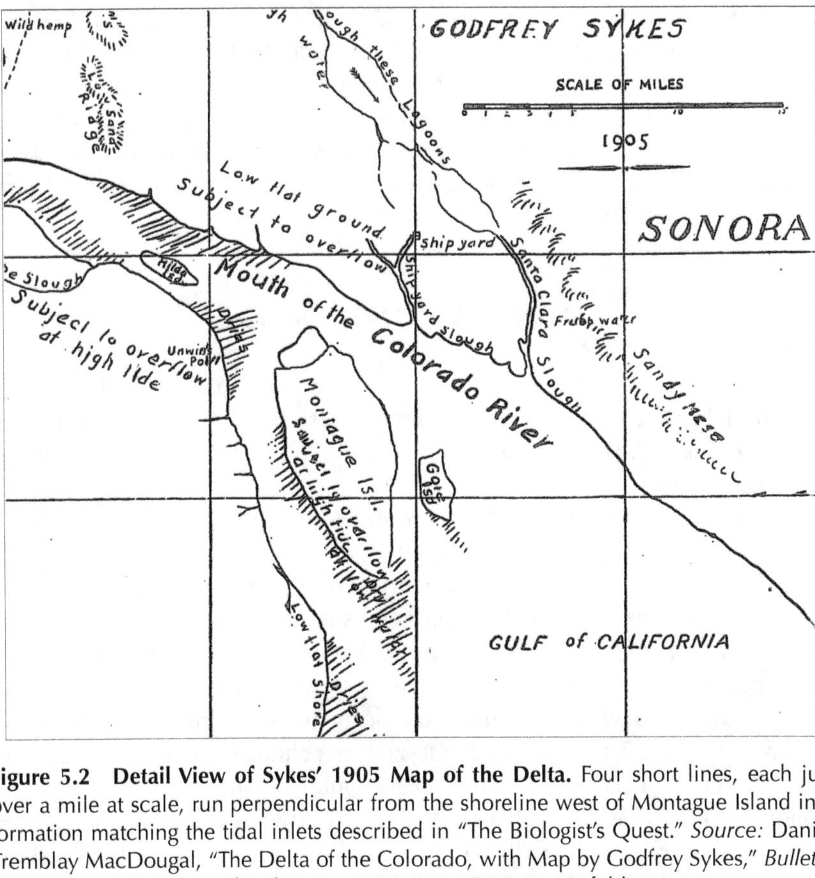

Figure 5.2 Detail View of Sykes' 1905 Map of the Delta. Four short lines, each just over a mile at scale, run perpendicular from the shoreline west of Montague Island in a formation matching the tidal inlets described in "The Biologist's Quest." *Source:* Daniel Tremblay MacDougal, "The Delta of the Colorado, with Map by Godfrey Sykes," *Bulletin of the American Geographical Society* 38.1 (Jan. 1906): 1–16, foldout.

between the elevation of the "quest" (with its implications of chivalry, though quixotic) and its object: not just a rat—the paradigmatic problem animal or pest, especially so for urban magazine subscribers—but one with a short tail.

This feature sets the species apart, taking the folk-biological vermin category of "rat" and converting it into life associated with Nature rather than with plague. But it also manages to depreciate it in ways that echo Kitti Quist's disposal of the snake, potentially evoking castration (destabilizing the masculinism so predominant in the history of natural history and in the mythology of frontiersmen like Kit Carson) and dismemberment (ruining the corporeal integrity of the scientific specimen). The short-tailed rat's body is an object whose value, in other words, is inseparable from its devaluation. This incongruity stands at odds with the common allegorical work of extinction narratives, which, Ursula K. Heise has written, often imagine that "part of one's national identity and culture might be preserved, revived, or changed for the better if an endangered species could be allowed to survive or an extinct one could be rediscovered."[31] "The Biologist's Quest" invokes this logic without obeying it, creating a borderlands ambiguation of the "national identity and culture" in question, and registering forms of precarity that do not break along the human-nonhuman line in the form of analogy. If this is an imaginative adaptation of material pulled from a local history of zoological knowledge, it is also an interpretation of the imaginative and material processes of such knowledge in the making.

MY NAME IS ORYZOMYS

> "He was walking by a brook one day, and saw a water-rat run past on the opposite bank in great haste. Almost immediately afterwards came a very fine stoat, hot in pursuit . . . and he expected to see the rat fall every moment. But such was not the case."
>
> —James Rodwell, *The Rat: Its History and Destructive Character*, 1858

In 1896, along the Río San José del Cabo, an estuarial river near Santa Anita at the southern tip of the Baja peninsula, Coolidge caught six specimens of an unfamiliar rice rat.[32] The specimens were sent on to the British Museum zoologist Oldfield Thomas. The following year, Thomas published a note about the new species under the designation *Oryzomys peninsulae*.[33] He wrote in a letter to Coolidge: "Oryzomys peninsulae is a particular surprise."[34] Another description of the species published in July 1901 (the same month as "The Biologist's Quest") lists the following characteristics: "Size rather large; ears rather small; tail short; color grayish; belly whitish; skull broad and massive."[35]

Tail short. This Baja swimmer sounds familiar. Coolidge's 1890s field-notes mention many, many rats—a daily entry might record having caught, say, seven kangaroo rats and two wood rats—but I have not found reference, in his journals or correspondence, to any species other than *O. peninsulae* that would fit the bill of being both short-tailed and semiaquatic. Given that this species was esteemed a "particular surprise" by zoological authorities, one could imagine that the story merited mentioning to a friend—perhaps, with a wry reflection or two, while passing the time on a transatlantic passage.

Of course, the identification I am insinuating—Coolidge's *O. peninsulae* as the rat that swims through the brackish sloughs of Lake's dreams—is unreasonable. This is not only because I'm unable to prove the absence of another similar species within the totality of Coolidge's victims. Nor is it because Coolidge found *O. peninsulae* farther south on the peninsula than Oskison's fictional biologist ever traveled, nor because McLean of the Pennsylvania Academy of Sciences denied this species' existence. This identification is, instead, unreasonable because "the Pennsylvania Academy of Sciences" did not exist either, not in the form imagined here, no more than Cooley or Kitti Quist did.[36] Because, that is, of the capacity of literature both to refer and to make, both to reproduce and also (and thereby) to speciate and split.[37] A fictional work generates taxonomies irreducible to those outside it.

Irreducible but not unconnected. If there aren't identities here—if Cooley isn't Coolidge; if these rats aren't rats—there are metonymies and resemblances, and they aren't confined to the immediate interpersonal channel (a college friendship) that brings us from source material to story. According to a recent review of the history of *O. peninsulae*'s classification by Michael D. Carleton and Joaquín Arroyo-Cabrales, only twenty-one specimens have ever been collected: six by Coolidge and fifteen more a decade later in 1906. Field teams tried to find more in 1979 and again in the early 1990s. Like Lake, they failed. All twenty-one specimens had been found along the Río San José. Largely dried up by irrigation, polluted by the infrastructural development of the tourism destination, the estuary has become an inhospitable habitat for *O. peninsulae*. As a result, the species is now probably critically endangered or extinct. But for several decades, beginning in 1971, nobody thought it existed at all. It was lumped, Professor McLean style, into the expanded description of another kind of *Oryzomys* (*couesi*), which lives across the gulf on the mainland. *O. peninsulae*'s taxonomic status as a distinct species has recently been reasserted, however, based on morphological observations of the extant specimens, as well as a phylogenetic hypothesis: its location in an isolated pocket around Cabo may have resulted from the tectonic rift that, about six million years ago, separated the peninsula from the mainland, creating the gulf and the conditions for a lineage-splitting event.[38] Rock strata and remains have brought the species back to recognition, if not to life (figure 5.3).

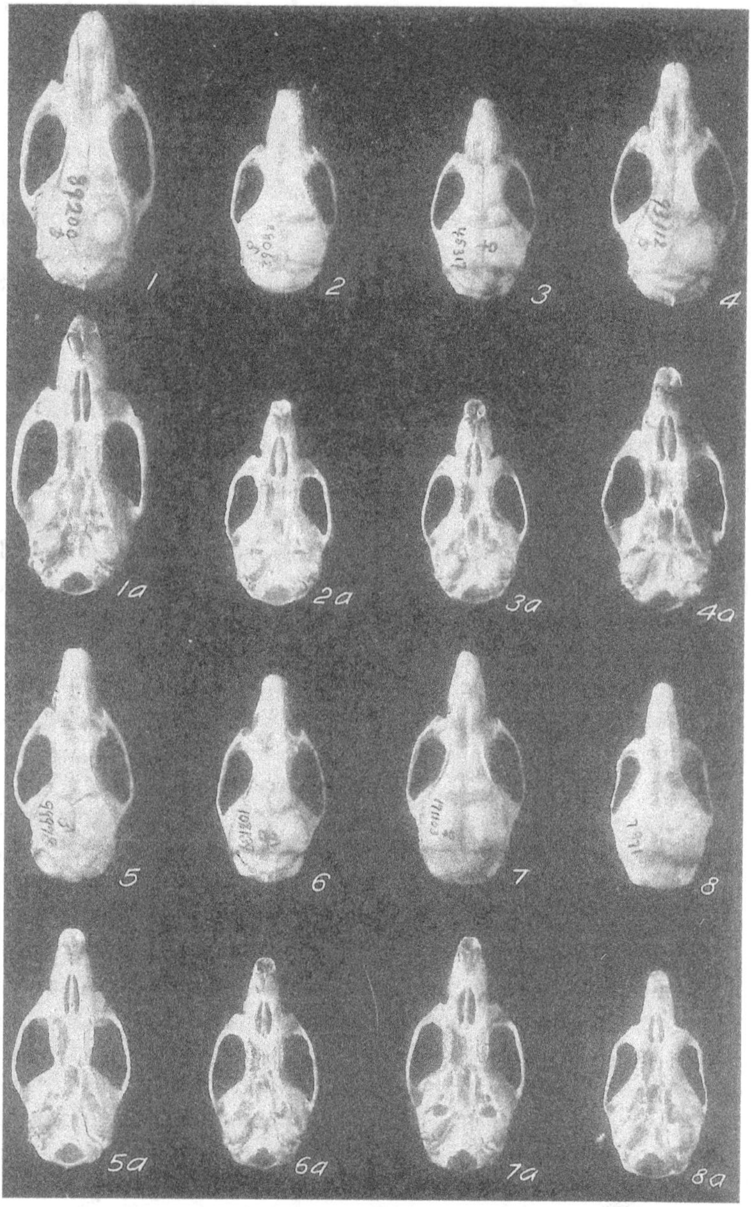

Figure 5.3 ***Oryzomys Peninsulae*** **at Bottom Right, 8 and 8a, and** ***O. Couesi*** **at 3 and 3a.** *Source:* "Skulls of Oryzomys" (Plate 1), in Edward Alphonso Goldman, *The Rice Rats of North America (Genus Oryzomys)*. U.S. Department of Agriculture, Bureau of Biological Survey: North American Fauna, no. 43 (Washington: Government Printing Office, 1918), facing p. 98.

At the other end of the peninsula from Cabo, what happened to *O. peninsulae*'s habitat happened to the Colorado.[39] It hasn't been possible for over half a century to get in a boat and follow Lake's itinerary, nor that of Aldo Leopold in 1922, who later recalled that, in the delta, the river "meandered in awesome jungles," "nowhere and everywhere."[40] By the 1960s "nowhere" was the key word: the Colorado had stopped flowing from Yuma to the gulf. Major causes include the construction of the Hoover Dam in the early 1930s—first authorized in 1929 by Coolidge's cousin Calvin, and then advanced under the presidency of Herbert Hoover, Geology major, Stanford class of 1895—which was followed, in the second half of the twentieth century, by Glen Canyon Dam and later by the Central Arizona Project aqueduct, among other water diversion projects.

But this process had already been accelerating since the turn of the century. Articles at that time describing enormous new irrigation projects in southwestern Arizona and California often compared the Colorado, in its potential, to the Nile. "The stupendous irrigation system is calculated to render the vast arid waste as fertile and useful as the delta of Egypt's sacred river," noted a representative April 1901 piece in the *Arizona Republican*.[41] In this comparison, agro-industrial development sounds like reflorescence. Here the "vision of the origin" so central to natural history in the period—a vision suffusing its institutional spaces, biopolitical discourses, and taxidermic practices—also finds expression in the geomorphic transformation of the zoological field itself.[42] The actual effects of that transformation, though, have involved severe desiccation and biodiversity loss throughout the borderlands. A 2012 initiative to revitalize the river and delta biomes, negotiated by the US and Mexican governments, led to strategic "pulse flows" of water released into the riverbed along with reductions in irrigation diversion.[43] To date they have had minimal lasting effect. A recent study of Cucapá responses to these environmental conditions notes that many people living around the delta articulate the exigencies of their situation, including conflicts over fishing access, not solely through the question of the water's disappearance but rather through the difficulty of finding work, "shift[ing] the terms of the debate onto the conditions of poverty that [make] feeding their families the ultimate priority."[44] And shifting, too, from a narrative of absence, of the river that vanished, to an assertion of presence.[45]

In turning from Oskison's narrative to its surprising reverberations across a longer and ongoing history of environmental violence—or perhaps they're unsurprising, perhaps this is the genre of such stories—I haven't meant to ascribe to him the kind of proleptic vision that Coolidge assumed over that little chipmunk's fate. The story of Lake's dehydrated blundering through the dunes and around the sloughs of the delta does not, for instance, anticipate the desiccation of the region. Nor do I want to overmeasure this short tale's moral or historical magnitude. That would, after all, entail falling right into the trap of its satire.

Still, "The Biologist's Quest" offers valuable provocations. Poaching on a network of naturalist, natural-historical, and ethnographic writing (a network that is both highly localized and transnational), the story unsettles the commonplaces of these overlapping fields. While riffing on a certain romance of fieldwork—lampooning the gendered identificatory space of this romance, if also probably reinforcing its allure in the reading experience of some *Overland Monthly* subscribers—the narrative also enfolds subjectivities, forms of labor, and temporalities that usually only haunt the margins of the kind of publications it plays on: the non-teleological temporalities of its spiral emplotment, for instance, but also of the infinite oscillations of a dream rat "swimming forever" back and forth as the tides ebb and flood. At the same time, and in part through its verbal oscillations—its puns, ironies, and nomenclatural slippages—the story develops a counternarrative to the favored plots of classificatory discourse. It makes weird taxa.

By doing so, it points up the situated contingency of classification systems, particularly as they take shape within contexts of corporeal and territorial expropriation. These contexts make ethical demands as urgent as the disappearance of the taxa such systems name. Among other responses to those demands, one might shift one's gaze from the chipmunk (or with the chipmunk; who knows, maybe it's still out there) to the mesh of materials and spatiotemporal structures—a fence, cotton of a certain grade, the workweek, a trap, a sense of plot—by which its life came to seem knowable and its death a sure thing.

NOTES

1. Dane Coolidge to Francis Coolidge and Bert Coolidge, 30 June 1895, Dane Coolidge Papers, Outgoing Correspondence.

2. On such temporal deixis as an "ecomimetic" blurring of distinctions between subject and environment, see Morton, *Ecology without Nature*, 3; in Coolidge's letter, though, this blur is a background against which animal death stands out in high definition.

3. See Barthes, *Camera Lucida*, 92. My comparison is also informed by Haraway's understanding of wildlife photography in "Teddy Bear Patriarchy."

4. This moment in Coolidge's letter has elements of the familiar "last of its kind" story, in which the death of the individual dramatizes that of the group or species; on the way that this convention "translates extinction into narrative," see Heise, *Imagining Extinction*, 38. Of course, this conceit would be undermined by the fact that "chipmunk" encompasses many different species designations, but here the term works as a vernacular species name or what philosophers of taxonomy would call a

folk-biological kind. The dynamic I'm describing in Coolidge's letter also involves what Bennett, thinking through the phenomenology of her own response to an assemblage of street debris including the body of a poisoned rat, describes as a shimmer between instrumentalized object and irreducible thing—an irreducibility embodied in the "singularity of *that* rat" (*Vibrant Matter*, 4).

5. On "future ghosts" see Morrill et al., "Before Dispossession," 3. On "perpetual vanishing," see N. Brown, "Logic of Settler Accumulation." On animal studies, Indigenous studies, and Oskison, see Hudson, "Domesticated Species."

6. Oskison, "A Letter to his Father: John Oskison Writes of his Visit in Europe," in *Tales*, 137–38.

7. John M. Oskison to Dane Coolidge, March 6, 1900, Dane Coolidge Papers, Incoming Correspondence. Coolidge's journal from the Europe trip lists specimens continuously numbered 1001–1458 (Dane Coolidge Field Book, 1900, Smithsonian Institution Archives, Acc. 12–232).

8. Dane Coolidge Field Book, 1897, Smithsonian Institution Archives, Acc. 12–433.

9. See Coolidge's field journal entry dated June 25, 1897 (Dane Coolidge Papers, Carton 1).

10. See Coolidge's field journal entry dated June 24, 1896 (Dane Coolidge Papers, Carton 1).

11. Oskison, "Biologist's Quest," 52.

12. Coolidge's 1896 Lower California journal (Dane Coolidge Papers, Carton 1) refers repeatedly to a José Maria, which may have informed Oskison's choice of this guide's name, although the character does not otherwise obviously resemble the person represented in Coolidge's journal.

13. Dane Coolidge to Francis Coolidge, July 15, 1900, Dane Coolidge Papers, Outgoing Correspondence.

14. "Dane Coolidge Returns." The Bronx Zoo, founded in 1895, was more properly "new" than Central Park in the 1890s. On the history of bringing rodents to East Coast parks—and on representations of class, ethnicity, and morality in the language of social reform that generally surrounded these programs—see Benson, "Urbanization of the Eastern Gray Squirrel." My work in this essay also benefits greatly from our many conversations on related questions and from his writing more generally.

15. The splitting or lumping of classifications was much debated in the period; see Theodore Roosevelt's criticisms of C. Hart Merriam's "overemphasis on minute points of variation" (*Letters*, 614).

16. Brooks, *Reading for the Plot*, 93.

17. On models of storying that open alternatives to "chrononormativity," see Rifkin, *Beyond Settler Time*, esp. 36–37.

18. See Haraway, "Teddy Bear Patriarchy," and Rony, "Taxidermy and Romantic Ethnography," in *Third Eye*, 99–126.

19. "Collecting Wild Animals."

20. There was, however, a Sociology department, where Mary Roberts Smith taught courses on race, immigration, and gender before founding the Sociology

program at Mills College. She married Coolidge in 1906. In 1939 they coauthored an ethnography titled *The Last of the Seris*. Despite that title, Seri people still live in Sonora and on the Gulf of California.

21. Miller, "Burial Customs," 288.

22. Justice's own answer is that Oskison develops a Beloved position that makes space for principles of Indian governance, although, he notes, Oskison's politics can be "inconsistent" (*Our Fire Survives the Storm*, 119).

23. Lake's name (which itself rounds out the story's proliferating bodies of water) seems to also get misspelled as "Jake" in the stylized drop-cap J that begins the story (52). Both this and the spellings of Quist/Quest may of course be typesetting errors, but the latter draws my particular attention because even in its standard form it evokes the wording of the title—because, in other words, whether or not one takes the alternate spelling to be accidental, the conditions of its probability and its suggestiveness are less likely to be so. My attention to both the sense of fluid futurity and the semantic play that surround this name is informed by the work of novelist, poet, and critic Gerald Vizenor, particularly by his account of "postindian warriors of survivance"—those who make of resignification a practice of ongoing survival, not as a mere biological fact but as an ongoing creative process (see Vizenor, "Ruins of Representation"). Although Vizenor has never in print made more than a brief mention of Oskison, my sense is that there are affinities in their understandings of the political potency of irony.

24. The responses to taxidermy depicted in "The Biologist's Quest" may have been informed by something that happened on Coolidge and Oskison's 1900 journey. The story opens with a reference to another biologist having been killed by a "superstitious Mexican" (52), which seems to alternatively nationalize some "superstitious peasants" Oskison remembered from the Europe trip: "Once the superstitious peasants threw stones at Dane until the priest came along and assured them that this American was only a harmless magic maker" (Oskison, "An Autobiographical Letter to Journalist Frederick S. Barde," in *Tales*, 139–43, 141).

25. Simpson, "On Ethnographic Refusal." In this volume, see also Figueroa, Perez, and Mantz on such refusals as "survival responses" and Heryford on Glissant's sense of "opacity." On naturalist guides' influence on zoological knowledge, see Jacobs, "Intimate Politics of Ornithology" (with thanks to Yuka Suzuki for pointing me to this piece). By suggesting that Kitti Quist's snake dance is "idiosyncratic," while writing, myself, as a white settler descended from European families with no affiliation with Indigenous nations of the Southwest (the Quecha/Yuma nation of which Kitti Quist is a fictional member; the Moqui/Hopi nation in which snake dancing is famously practiced), I claim no knowledge of snake dance practices beyond my awareness of a textual record produced by outsiders whom I do not take to be authoritative on the topic. As my reading here of Kitti Quist's dance suggests, my sense—perhaps wrong, given these limitations—is that this imagined performance is, in key respects, an anomalous one through whose peculiarities Oskison, probably not closely familiar with snake dancing himself at this point in his life, develops a critique of scientific (including anthropological) paradigms of representation. In this reading, taking Kitti Quist's dance as a canny comment, I

diverge from Cox's description of the same moment as a masculinist display that reveals the guide's "desperation." More generally, Cox finds in "The Biologist's Quest" a stereotype of Mexican political disorder—"Mexico promises death rather than a new life"—that is reinforced in its reference to Joe Maria as "lazy" ("Learn to Talk Yaqui," 405). This analysis importantly reframes Oskison's politics in a transnational context and illustrates his capacity for expressions of bigotry and American expansionism. But "The Biologist's Quest" mocks rather than advances these attitudes. It associates Mexico with death in the perspective of Lake, a fool, and in that of his East Coast employers, at their conspicuously ill-informed remove. The guide's ostensible laziness, meanwhile, is contradicted by his obvious competence, and the epithet is used precisely when it is least appropriate, not only in that Joe Maria's plan saves Lake's life but also in that he is described in the same breath as acting "frantically" (56). For another approach to the rhetorical complexities of Oskison's management of the perspective of white characters, see Hunnef, "Alternative Histories."

26. Coolidge, "Gentle Art," 676.

27. In this way the snake refuses to become the kind of "boundary object" that the sought-after rat exemplifies; see Star and Griesemer, "Institutional Ecology." In a new-materialist register, one might say that Kitti Quist's destruction of the snake enacts a violent reminder of "a culture of things irreducible to the culture of objects" (Bennett, *Vibrant Matter*, 5).

28. Fisher, "William Wightman Price," 56. See also Price's own brief account, "Some Winter Birds." On the 1895 trip, see correspondence from Dane Coolidge to Bert Coolidge, May 12, 1895, Dane Coolidge Papers, Outgoing Correspondence.

29. In this dream-state encounter with a biological specimen, Oskison elaborates on a scenario he had explored in one of his earliest published stories, "A Laboratory Fancy," which involves a student who falls asleep in the laboratory and dreams that the protozoon he is studying starts to speak to him about the violence of biological study. The ambivalence with which Oskison treats the dream-space in these texts, as a site of both violent alienation and empathetic experience, also features in Vizenor's autobiographical account of a terrible hunt. He shoots a squirrel but fails to kill it immediately; the extended description of what follows combines brutal physicality and fantastic identification. The squirrel tries "to escape from my dream, the city in me," Vizenor writes; "I understood his instinct to escape; in a dream we reached up with our right paw, shattered and blood soaked, but it was not there" ("October 1957: Death Song to a Red Rodent," in *Interior Landscapes*, 167–70, 168).

30. I have not found examples of the precise phrasing "vision quest" in print prior to Oskison's story, but it does come into academic discourse around this time (before coming into wider usage in the mid-twentieth century, with the commoditized representation of "vision quests" and "spirit animals" in popular settler culture). In 1882 Alice C. Fletcher describes a sacred fasting ceremony as a "quest for the raven or the stone" in which a young man "may see in his vision one of these symbols" ("Religious Ceremony," 289). By the time Robert Lowie uses the phrase in 1914, he refers to the "familiar heading" of the "vision quest" ("Ceremonialism," 627).

31. Heise, *Imagining Extinction*, 49.

32. Coolidge had been hired by Price to lead "a party to southern Lower California to collect birds and mammals" (Nelson, *Lower California*, 145).

33. Thomas, "Description of New Bats and Rodents," 544–53, description of *O. peninsulae* at 548–49.

34. British Museum [signed Oldfield Thomas] to Dane Coolidge, March 3, 1898, Dane Coolidge Papers, Incoming Correspondence.

35. Merriam, "Synopsis," 278.

36. The organization occasionally referred to, in the period, as the "Pennsylvania Scientific Society" was a scientific fraternity at Penn that hosted a lecture series; a more probable institution for the activities Oskison mentions (collecting specimens, publishing reports) would have been the Academy of Natural Sciences of Philadelphia.

37. On the history of the idea of the "monstrous" as encompassing not only surprising hybrids but also classifications that "had their genesis in human ingenuity, imagination, or violence," see Ritvo, *Platypus and the Mermaid*, 133.

38. Carleton and Arroyo-Cabrales, "Review of the Oryzomys couesi Complex," 114–15.

39. See, in this volume, both Ottum on Lydia Millet's fictional version of a similar scenario involving the impact of development on kangaroo rats, and Oele in response to Michel Serres on the weakening and death of rivers.

40. Leopold, "Green Lagoons," 150. Lamenting the agricultural despoiling of an untouched Edenic space that ostensibly has "no place names"—a "milk-and-honey wilderness" where journeyers find themselves "back in the Pleistocene"—the essay culminates in a lyrical critique whose premises include Indigenous erasure (156, 155, 157).

41. "Uncle Sam's Egypt and Nile," 6.

42. Haraway, "Teddy Bear Patriarchy," 20. See also Yusoff, "Anthropogenesis."

43. The project is known as the Minute 319 project, in reference to an element of the 1944 water treaty, and was renewed in 2017 as the Minute 323 agreement.

44. Muehlmann, *Where the River Ends*, 5.

45. This assertion also refuses a discursive history in which anthropological predictions of Cucapá disappearance were entwined with agricultural claims of the Colorado's potential to support an increasing settler population; this claim is advanced in explicitly social-evolutionary terms (and in direct comparison with the Nile) in what is perhaps the period's most detailed geographical study of the area: MacDougal, "Delta" (see 15–16).

BIBLIOGRAPHY

Barthes, Roland. *Camera Lucida: Reflections on Photography*, trans. Richard Howard (New York: Macmillan, 1981).

Bennett, Jane. *Vibrant Matter: A Political Ecology of Things* (Durham: Duke University Press, 2010).

Benson, Etienne. "The Urbanization of the Eastern Gray Squirrel in the United States." *Journal of American History* 100.3 (December 2013): 691–710.

Brooks, Peter. *Reading for the Plot: Design and Intention in Narrative* (New York: Knopf, 1984).

Brown, Kirby. *Stoking the Fire: Nationhood in Cherokee Writing, 1907–1970* (Norman: Oklahoma University Press, 2018).

Brown, Nicholas A. "The Logic of Settler Accumulation in a Landscape of Perpetual Vanishing." *Settler Colonial Studies* 4.1 (2013): 1–26.

Carleton, Michael D. and Joaquín Arroyo-Cabrales. "Review of the Oryzomys couesi Complex (Rodentia: Cricetidae: Sigmodontinae) in Western Mexico." *Bulletin of the American Museum of Natural History* 331 (Dec. 2009): 94–127.

"Collecting Wild Animals." *Daily Palo Alto*, 31 October 1899, 1.

Coolidge, Dane. "The Gentle Art of Photographing Rattlesnakes." *Metropolitan* 28.6 (Sept. 1908): 673–78.

Coolidge, Dane and Mary R. Coolidge. *The Last of the Seris* (New York: E. P. Dutton and Co., 1939).

Cox, James H. "'Learn to Talk Yaqui': Mexico and the Cherokee Politics of John Milton Oskison and Will Rogers." *Western American Literature* 48.4 (Winter 2014): 400–21.

Dane Coolidge Papers, BANC MSS C-H 82, Bancroft Library, University of California, Berkeley.

"Dane Coolidge Returns." *Daily Palo Alto*, 26 April 1899, 1.

Fisher, Walter J. "William Wightman Price." *The Condor* 25.2 (Mar.–Apr. 1923): 50–57.

Fletcher, Alice C. "The Religious Ceremony of The Four Winds or Quarters, as Observed by the Santee Sioux." In *Reports of the Peabody Museum of Archaeology and Ethnology*, vol. 3 (Cambridge, MA: Harvard University Press, 1887), 289–95.

Haraway, Donna. "Teddy Bear Patriarchy: Taxidermy in the Garden of Eden, New York City, 1908–1936," *Social Text* 11 (Winter 1984–95): 20–64.

Heise, Ursula K. *Imagining Extinction: The Cultural Meanings of Endangered Species* (Chicago: University of Chicago Press, 2016).

Hudson, Brian K. "Domesticated Species in D'Arcy McNickle's *The Surrounded* and John M. Oskison's *Brothers Three*." *Studies in American Indian Literatures* 28.2 (Summer 2016): 80–108.

Hunnef, Jenna. "Alternative Histories of the Old Indian Territory: John Milton Oskison's Outlaw Hypotheses." *Western American Literature* 53.3 (Fall 2018): 339–71.

Jacobs, Nancy J. "The Intimate Politics of Ornithology in Colonial Africa." *Comparative Studies in Society and History* 48.3 (July 2006): 564–603.

Justice, Daniel Heath. *Our Fire Survives the Storm: A Cherokee Literary History* (Minneapolis: University of Minnesota Press, 2006).

Leopold, Aldo. "The Green Lagoons." In *Sand County Almanac with Essays on Conservation from Round River* (New York: Oxford University Press, 1966): 150–8.

Lowie, Robert H. "Ceremonialism in North America." *American Anthropologist* 16.4 (1914): 602–31.

MacDougal, Daniel Tremblay. "The Delta of the Colorado, with Map by Godrey Sykes." *Bulletin of the American Geographical Society* 38.1 (Jan. 1906): 1–16.

Merriam, C. Hart. "Synopsis of the Rice Rats (Genus Oryzomys) of the United States and Mexico." *Proceedings of the Washington Academy of Sciences*, vol. 3 (26 July 1901): 273–95.

Miller, Mabel L. "Burial Customs and Ideas Among the Indians of the Sacramento Valley." *Sequoia*, 15 Mar. 1895, 288–9.

Morrill, Angie, Eve Tuck, and the Super Futures Haunt Qollective. "Before Dispossession, or Surviving It," *Liminalities: A Journal of Performance Studies* 12.1 (2016): 2–20.

Morton, Timothy. *Ecology without Nature: Rethinking Environmental Aesthetics* (Cambridge, MA: Harvard University Press, 2007).

Muehlmann, Shaylih. *Where the River Ends: Contested Indigeneity in the Mexican Colorado Delta* (Durham: Duke University Press, 2013).

Nelson, Edward W. *Lower California and Its Natural Resources*, Memoirs of the National Academy of Sciences (Washington, DC: Government Printing Office, 1921).

Oskison, John M. "The Biologist's Quest," *Overland Monthly*, July 1901, 52–7.

———. *The Brothers Three* (New York: Macmillan, 1935).

———. "A Laboratory Fancy," *Sequoia*, 5 Feb. 1897, 213–14.

———. *Tales of the Old Indian Territory and Essays on the Indian Condition*, ed. Lionel Larré (Lincoln: University of Nebraska Press, 2012).

Price, W. W. "Some Winter Birds of the Lower Colorado Valley," *Bulletin of the Cooper Ornithological Club* 1.5 (Sep.–Oct. 1899): 89–93.

Rifkin, Mark. *Beyond Settler Time: Temporal Sovereignty and Indigenous Self-Determination* (Durham: Duke University Press, 2017).

Ritvo, Harriet. *The Platypus and the Mermaid and Other Figments of the Classifying Imagination* (Cambridge, MA: Harvard University Press, 1997).

Rony, Fatimah Tobing. *The Third Eye: Race, Cinema, and Ethnographic Spectacle* (Durham: Duke University Press, 1996).

Roosevelt, Theodore. *Letters*, ed. Elting E. Morison, vol. 1 (Cambridge, MA: Harvard University Press, 1951).

Simpson, Audra. "On Ethnographic Refusal: Indigeneity, 'Voice' and Colonial Citizenship." *Junctures: The Journal for Thematic Dialogue* 9 (Dec. 2007): 67–80.

Smithsonian Institution Archives, Field Book Project 1855–2010.

Star, Susan Leigh and James Griesemer. "Institutional Ecology, 'Translations,' and Boundary Objects: Amateurs and Professionals in Berkeley's Museum of Vertebrate Zoology, 1907–39," *Social Studies of Science* 19.3 (1989): 387–420.

"Uncle Sam's Egypt and Nile: Vast Desert around Colorado River Delta to Bloom." *Arizona Republican* (Phoenix, AZ), 29 April 1901, 6.

Vizenor, Gerald. *Interior Landscapes: Autobiographical Myths and Metaphors* (Albany: State University of New York Press, 2009).

———. "The Ruins of Representation: Shadow Survivance and the Literature of Dominance." *American Indian Quarterly* 17.1 (Winter 1993): 7–30.

Yusoff, Kathryn. "Anthropogenesis: Origins and Endings in the Anthropocene." *Theory, Culture and Society* 33.2 (March 2016): 3–28.

Chapter 6

Contesting *Extinction* through a Praxis of Language Reclamation

Wesley Y. Leonard

I begin this essay by locating myself in a community narrative called *myaamiaki eemamwiciki* (Miami awakening), which centers the story of how my tribe brought our language, which was once labeled "extinct," back into our community as part of a larger cultural renaissance and political resurgence. I am a citizen of the Miami Tribe of Oklahoma who became a linguist to support my tribal community's efforts to reclaim our language, which in English is also called Miami but in our language is described as *myaamiaataweenki*, a verb that means "myaamia is spoken." However, many tribal members use the word that describes our peoplehood, *myaamia*, to also refer to the language, a convention I adopt for this chapter.[1]

Stemming from a long history of subjugation as a result of colonization, settler colonialism, and a host of related *-isms*, intergenerational transmission of myaamia started to decrease at some point in the late nineteenth century. This history includes two forced removals by the US government of a large segment of the Miami community, including my direct ancestors, from our original homelands in Indiana and Ohio, first to Kansas and then later to Indian Territory (now Oklahoma—hence the name of my tribal nation). It includes the imposition of an ideology that Miami ways were savage and unfit for modern life, an idea that has been imposed on Native American cultures more generally. This history includes my ancestors' experiences in federal Indian boarding schools, where they, according to one of my family elders, "weren't allowed to talk Indian."

Without intergenerational transmission, myaamia eventually ceased to be actively known almost entirely at some point around the 1960s, perhaps a little later, with the passing of the final group of people who had grown up with the language prior to its period of silence. This was the point at which myaamia became "extinct" according to linguists. However, myaamia remained well described in close to 300 years of written documentation held in archives.[2] Linguistic research using that documentation began in the late 1980s, and after this came grassroots language recovery efforts in the early 1990s. These early community-driven efforts were followed in the late 1990s, and more significantly in the 2000s, by programs formally sponsored by our tribal government such as language camps and workshops.[3]

My role in myaamiaki eemamwiciki began in my early adulthood in the mid 1990s, initially as a language learner and later as a professional linguist specializing in what I describe as *Indigenous language work*, which I adopt as an umbrella term for Indigenous language documentation, description, analysis, learning, teaching, and advocacy. However, the connection and commitment I feel to my community was cultivated in my childhood through tribal activities and relationships with Miami places and people. Especially important were the interactions I had with my late grandfather, who served as our tribal chief for many years and who committed his life to strengthening our tribal community. An educator himself, he was particularly keen about the role of education in addressing tribal needs, and by extension, about the possibilities for research to support community goals such as the restoration of our language.

The research supported by my grandfather and other tribal leaders has been effective. As a result of this work, there are many myaamia learners with varying levels of proficiency. Our language is actively used in several Miami families, shows up in tribal publications such as our tribal newspaper, and marks tribal places through written myaamia signage. The basic story is that a group of people started speaking and teaching myaamia while concurrently learning it from the vast corpus of written records, which contain dictionaries, recorded narratives, grammatical information, and descriptions of usage norms. Linguistic analysis, particularly through consultation of related Algonquian languages, continues to facilitate filling in gaps in the records. Our language also continues to evolve in response to our changing needs and environments, particularly through words for newer items like *kiinteelintaakani* 'computer' (literally, "the thing that thinks fast"). In summary, Miami people are learning and speaking myaamia in a way that reflects our contemporary community. Though English remains the primary language within most community contexts, myaamia is increasingly common. It embeds our unique tribal culture and connects us as a people in a way that English cannot.

In other words, myaamia is not extinct, and never was. However, this sort of logical explanation does not align with the broader set of logics and received "truths" about Indigenous communities and languages that are promoted by wider society. Of special importance to this chapter are the "colonial epistemological invasions"[4] that normalize the "extinctions" of Native American languages as a natural outcome, while also greatly curtailing "authenticity" by comparing contemporary Native American languages and language users to an imagined baseline anchored in historical ethnographic accounts by non-Indigenous agents. Following earlier work, I refer to these interrelated logics, and to their operationalization in descriptions of Indigenous languages and language work, as the "dominant discourse."[5]

According to the dominant discourse, speaking a Native American language that has been brought back into use after a full break in intergenerational transmission does not count as evidence of the language's vitality because the current forms of speech are not considered legitimate. By the same token, those who speak myaamia today are often not considered "real" speakers for purposes of assessing the status of myaamia. Even after a quarter century of active myaamia use, *Wikipedia* continues to label myaamia as extinct.[6] Conversely, members of my community exercise our sovereignty by eschewing what we increasingly refer to as the "*e*-word" and instead adopting the term sleeping to describe our language as it existed during its dormancy; by extension, myaamia is now a formerly sleeping language.[7] Still, colonial agents attempt to maintain power. This is exemplified by how somebody (more likely non-Indigenous somebod*ies*) always reverses the corrections I make to the *Wikipedia* entry for my community's language.

Arising from the experiences described above, and especially by observing several Miami language leaders' insistence on guiding language work on our own tribal terms, I have been developing a framework of *language reclamation* to describe, theorize, and build further capacity for the tribally centered language restoration practices I have observed in my own community and others.[8] Language reclamation identifies and addresses the underlying issues that precipitate language shift in a given community, and centers community goals and ways of knowing in all areas of language work. Reclamation praxis by nature is transformative in that it actively dismantles the underlying structures and ideologies, along with their ensuing practices, which make eventualities such as myaamiaki eemamwiciki unexpected to those who promote the dominant discourse.

I differentiate language reclamation, which is anchored in Indigenous needs and ways of knowing, from efforts that privilege non-Indigenous approaches to language work. These dominant approaches often engender a focus on direct linguistic measures such as increasing the number of speakers. From this dominant lens, "success" is assumed to occur when the transmission and usage patterns of the Indigenous language, along with the community's ideas about what language is meant to be and do, mirror those of wider society. To describe this latter approach, I use the more common term *language revitalization*, following the idea that this term suggests that language is the unit that needs to be fixed.[9] As recognized by Chickasaw education scholar Kari A. B. Chew, Indigenous languages "within themselves, are already vital . . .

and cannot be conceived of as separate from the communities that claim and speak them." Reclamation thus places the focus on how community members return to the practice of using their languages "as a means to strengthen cultural identity and to resist hegemonic legacies of colonization."[10]

While a common language goal is to restore language use and transmission patterns in reference to a community's historical norm, reclamation is not about recreating the past. Rather, it is about identifying and actively fostering "the appropriate cultural context and sense of value that the language would likely have always had if not for colonization."[11] As a decolonial intervention that starts from community histories and needs, language reclamation emphasizes the factors that instigate language shift such as removal and land dispossession, colonial education, and economic subjugation. It also recognizes how the original causes of language shift often remain today, and thus must be identified and addressed.

My goal in the remainder of this chapter is to share key tenets of a praxis of language reclamation by illustrating how decolonial Indigenous logics, grounded in narratives of resurgence such as that of my community, provide alternatives to the logics of colonization and settler colonialism that underlie and reproduce the dominant discourse. As I have discussed above, I will privilege the specific narrative of myaamiaki eemamwiciki to disrupt one of the worst facets of settler colonialism—that tribally specific stories are suppressed as part of a racialization process that lumps diverse Original Peoples into a collective unit, hence erasing our discrete nationhood. I share myaamiaki eemamwiciki with the hope that doing so will support other communities. However, I follow Michi Saagiig Nishnaabeg storyteller, scholar, and activist Leanne Betasamosake Simpson in emphasizing that trajectories of Indigenous resurgence, of which language reclamation is one type, should never be prescriptive.[12] As with Indigenous approaches in general, reclamation is ultimately a local phenomenon; its details emerge in particular histories, geographies, and epistemologies, and its operationalization is an articulation of sovereignty. Out of respect for the diversity of thought within the Miami community, I further emphasize that the commentary in this chapter reflects my personal views, which developed in a context of relationships with other Miamis but are ultimately my own. I begin with the broader response to "endangered languages" that motivate many of my ideas about reclamation.

COLONIALISM IN LINGUISTICS AND IN THE ENDANGERED LANGUAGES MOVEMENT

To understand language shift in communities such as my own, and ultimately to be able to respond through a praxis of reclamation, it is useful to consider how general tenets of settler colonialism inform the specific situation of widespread Indigenous language shift, and how different stakeholders respond. Of special importance is the field of Linguistics.[13] Although there have long been efforts within Linguistics to theorize and address minority language shift, many linguists point to a 1992 "Endangered Languages" special series of essays published in *Language*, the flagship journal of the Linguistic Society of America, as the major turning point where this became a significant topic within the field. Particularly referenced within this collection is "The World's Languages in Crisis," which paints a bleak picture and claims that "[l]anguages no longer being learned as mother-tongue by children are beyond mere endangerment, for, unless the course is somehow dramatically reversed, they are already doomed to extinction, like species lacking reproductive capacity."[14] I appreciate that this statement at least raises the possibility of alternative futures, but as I elaborate on below, it is problematic in that it normalizes assumptions from the dominant discourse, and frames Indigenous languages as biological species.[15]

After more than two decades of involvement in Miami language efforts, I chuckle but also shake my head in dismay as I remember the reactions I used to receive as a Linguistics student when I told people about how Miamis were working to bring myaamia back into our community from documentation. One recurring response was that this was a nice idea, but that it was not going to work. However, other linguists, particularly senior scholars of Native American languages, left off the first part, just directly saying my community's efforts were not going to work and even mocking these goals.[16] I came to recognize that such attempts to eliminate people like me from articulating Indigenous alternatives represented their attempt to maintain power over Indigenous languages. Indeed, what Wolfe describes as the "logic of elimination"[17] as a strategy to maintain territory is "embedded into every aspect of the settler colonial structures and its disciplines"[18]—and Linguistics is a colonial discipline.[19]

The colonial logics embedded into Linguistics are revealed when linguists lament language shift as an existential problem for Linguistics. Since theories about language cannot be complete without data from diverse languages, there is a significant push for documenting and describing endangered languages—a category that includes the majority of the world's Indigenous languages. Some of the resulting products support community language work, but the underlying premise stated above represents a colonial norm of exploiting Indigenous communities' resources. At the same time, however, there is in Linguistics a growing social justice movement with accompanying calls for supporting Indigenous communities through the development of community-oriented products, such as materials for teaching Indigenous languages. More broadly, sociopolitical concerns about Indigenous languages are increasingly being considered within academia as well as by governments, funding agencies, and non-profit organizations. A prominent example of global scale is the United Nations General Assembly's declaration of 2019 as the International Year of Indigenous Languages.

The collective response to recent patterns of Indigenous language shift has had many effects, several of which I deem to be positive—increased awareness, funding, research, and language programs, including many run by Indigenous communities. My concern, however, is that even the social justice-oriented work that is becoming more common exhibits a pervasive pattern of responding to Indigenous language shift through colonial assumptions, rhetorics, and practices that draw heavily from the dominant discourse rather than reclamation praxis. Especially common is the adoption of endangerment as a metaphor to describe the pattern of language shift as well as to guide the response, which might be described as the Endangered Languages Movement. What follows is my critical summary of the Endangered Languages Movement, with italicization in the remainder of this section marking important recurring phrases.

The key issue that underlies the Endangered Languages Movement is true: There has been a very significant decline in the use and transmission of minority languages. Though this pattern is global, the decline has been especially extreme in North America, the context of most of my experience. *While languages have always died*, according to the Movement, *the recent scope of death is unprecedented.* The main reason this is occurring is easy to identify: Parallel to how the Anthropocene is the anchor for the Sixth Extinction of biological species, the major cause of recent Indigenous language shift is the influx and destruction associated with what Sylvia Wynter describes as the "overrepresentation of Man." In other words, language shift occurs in contexts of dominant groups (Man) subjugating the Other, who are not considered Man since their humanity is erased.[20] US policies to rupture Native American communities and languages accelerated in the nineteenth century, which is when children stopped learning myaamia. myaamia then went out of use completely with the passing of those individuals.

However, although a growing body of critical scholarship engages with these power dynamics and identifies settler colonialism and related structures as root causes of the contemporary situation,[21] the public face of the Endangered Languages Movement is limited in its engagement with these ideas. Community members' negative language attitudes and vague sociopolitical variables such as *language shame* or *economic factors* are sometimes mentioned, but without additional probing as to why people might have developed an "ideology of contempt"[22] about their own languages, or how economic disparities develop and are maintained in the first place.

The erasure of colonial agency, a rhetorical strategy described by Chickasaw linguist Jenny L. Davis in her critique of the Endangered

Languages Movement, is common in the dominant discourse.[23] As Davis notes, this strategy masks the colonial violence that begets Native American language endangerment and dormancy, often misattributing agency onto the victims: communities *lose* their languages because *they stop speaking them* or *choose not to pass them on*. (But under what circumstances would somebody make such a choice?) Another insidious example draws from social Darwinism to falsely locate the problem in the Indigenous languages themselves; language shift is deemed a natural outcome of modernity under the unfounded argument that these languages must be poorly suited for contemporary communication needs. And in other cases, the cause is simply unaddressed. Especially common are variants of the phrase *Indigenous languages are disappearing*. For example, the main page of the website for the United Nations' International Year of Indigenous Languages adopts this strategy, detailing the "immense value" of Indigenous languages but providing no context as to why "languages around the world continue to disappear at an alarming rate."[24]

Disappear, along with synonyms such as *vanish*, is a special type of verb, an unaccusative, defined as one for which the grammatical subject is not the agent or doer, but rather functions as the recipient or experiencer akin to a prototypical direct object. Using unaccusative verbs to describe language shift masks a key variable that language reclamation must center—the cause. A praxis of reclamation calls for descriptions such as the following: "[explicitly named agent] disappears languages." In fact, Indigenous communities' shift away from their original languages invariably involve power imbalances, often with extreme subjugation of the language community and control of the community's resources by dominant groups.

It is common in the Movement to enumerate the looming loss of language diversity as an outcome of current trends, noting, for example, that there are estimated to be around 7,000 languages in the world today, and that *up to 90%* (specific estimates vary) of those *will be extinct within a hundred years*. Especially frequent, and repeated throughout academic work as well as popular media, is the "truth" that *a language dies every two weeks*.[25] A common response involves calling attention to the assumed plight of Indigenous and other minority languages, framing language shift as a unidirectional phenomenon characterized by increasingly severe stages of endangerment such as *threatened*, *moribund*, or *nearly extinct*. The endpoint is *extinction*—an eventuality already reached by languages like myaamia, according to the dominant discourse.

Terms like "extinction" can be used intentionally as tools of subjugation, and I have occasionally observed this sort of malicious intent. However, it is more common for the Endangered Languages Movement to adopt a benevolent tone. Nevertheless, the frame of "endangered languages" still inhibits reclamation. The metaphor carries a clear teleological trajectory toward extinction, thus obscuring the alternative futures that are crucial to reclamation praxis. Reaching the expected endpoint of extinction, in turn, renders language restoration ontologically impossible. Not surprisingly, languages that carry this label are usually beyond the calls for action that characterize the Endangered Languages Movement.

Instead, the Movement centers the looming loss for humanity, often operationalized within an assumption of universal ownership of languages in ways that disregard Indigenous communities' sovereignty.[26] The risk of *losing our heritage* (where "our" refers to dominant society) is offered as a reason to save those languages that still have speakers—in this context where speakerhood is narrowly defined to include only fluent, native speakers.[27] The concept of "saving" does include support for some revitalization efforts, particularly those involving methods that align with normative models of language learning. Even if ostensibly meant to support Indigenous communities, however, such dominant approaches easily further the colonial project by controlling how members of Indigenous communities relate to our own languages.

Beyond the rhetorical and ideological limitations imposed by discourses of language endangerment, reclamation is further inhibited because the Endangered Languages Movement also promotes what Comanche scholar Barbra A. Meek describes as a "discourse of failure" surrounding Native American language efforts: "Expectations tend to assume a status quo defined around failure, the result of some innate limitation on the part of Indian people. Success is written off as an anomaly."[28] There are, however, some success stories within the Endangered Languages Movement, and these get touted as special examples. The relatively few efforts that are deemed successful, for endangered biological species and endangered languages alike, are those characterized by an increase in numbers (individuals of the species and language speakers, respectively). However, while there is an increasing recognition, even among the general non-Indigenous public, that reversing the decline of biological species lies in protecting or recreating their appropriate ecological habitats, the parallel thinking for Indigenous languages—that language reclamation entails a resurgence of physical, spiritual, and relational contexts that support language learning and use—remains relatively uncommon outside of academic literature.

This example of ignoring how languages exist in complex environments exemplifies linguistic extraction, a rhetorical strategy described as follows by Davis:

> defining, analysing, and representing languages and the people connected to them separately from the complex socio-historical, political, and deeply

> personal contexts in which they actually occur. Linguistic extraction, then, renders languages into extractable objects that can be collected, preserved, utilised, and even admired. Critically, linguistic extraction is not solely the collection of endangered and Indigenous languages in ways that often render them inaccessible to their communities, but also the presentation of languages as objects, or data, without their complex and varied human contexts. . . . This literal and metaphorical extraction from context is itself a colonial enterprise and often a cornerstone of Western science—one that removes people from homelands, loots objects from graves in the name of science and education, and disassociates products from those who labour to produce them. In other words, it celebrates the *empire* in *empirical*.[29]

Beyond the idea that linguistic extraction inhibits language reclamation, the notion raised by Davis of languages being presented as objects is also very important. It is common in Linguistics to conceive of languages as discrete lexicogrammatical objects—specific sets of grammatical patterns and vocabulary—which are then rendered into "data" for theoretical examination using structural units such as sounds (Phonetics and Phonology), words (Morphology), and clauses (Syntax). In other words, Indigenous languages are extracted from their original contexts and reproduced as empirical scientific objects. This obscures how language as a social practice draws meaning from the ways in which it "is embedded in a broader cultural matrix," and that changes in how people use language occur because something has changed within that broader cultural matrix.[30] Though some linguistic research specifically focuses on social issues, an ongoing limitation even within the latter body of work is that much of it bolsters academic views of language by reducing Indigenous community beliefs to "ideologies" that can be reported on and analyzed within dominant theories, rather than operationalized as theories from which new knowledge is produced.

ERASING INVASIVE CONCEPTS, RESTORING INDIGENOUS CONCEPTS

By drawing attention to the full cycle within erase-and-replace processes of settler colonialism, reclamation goes beyond addressing the initial erasure to also interrogate the nature of the replacement. Reclamation has an erase-and-restore logic, where the target of erasure is subjugation and the associated invasive concepts, and successful restoration is that which meets community goals. Drawing from Davis's arguments about the erasure of colonial agency and the related process of linguistic extraction, in this section I detail a central goal of reclamation in the context of myaamiaki eemamwiciki: restoring relationships.

Reclamation praxis firmly eschews ideas of languages as extractable objects, instead focusing on the intersections of languages and communities' particular histories, needs, and epistemologies. By extension, language work, just as it must center the goals of a particular language community, must also privilege that community's conceptual views of language and related worldviews. myaamiaki eemamwiciki provides an illustrative example that addresses the general issue of centering community definitions as well as the specific example focused on in this chapter of contesting extinction.

Though I have thus far been writing about language as a noun, owing to the grammar of English which fosters this objectification, it is important to stress that the full word Miamis use to describe our language, *myaamiaataweenki*, is actually a verb whose meaning is closer to "myaamia is spoken."[31] That is, *myaamiaataweenki* describes *languaging*—a word used in critical approaches to Applied Linguistics to capture language-as-relational-interaction rather than language-as-object. Referencing this grammar, Miami language reclamation leader Daryl Baldwin observes that concepts like *death* and *extinction* "can really only be applied to entities (animate or inanimate) and not to actions. Speaking a language is not a thing, and therefore it cannot die . . . it can stop being spoken [but] there really isn't any notion of permanence implied here."[32] Indeed, myaamia ceased to be spoken—but then members of my community started speaking it again. During a conversation with me, myaamia language teacher Jarrid Baldwin, son of Daryl Baldwin, defined myaamia as "how a community connects to each other and how they express . . . themselves and their culture to each other," hence embedding the Miami people and our relational interactions into the definition.[33]

Summer language camps for Miami youth illustrate the critical relationality that guides Miami languaging.[34] These camps support reclamation by situating language into relational contexts, particularly those that have been weakened through processes of colonization and settler colonialism. For instance, as Miami culture is ecologically based, myaamia education often occurs specifically in relation to what is happening in the natural environment and participants directly engage with the landscape in numerous ways. Planting, gathering, and observing the environment are common activities, and the interconnections of culture and place are built into camp themes. For example, one of the rotating camp themes is *miiwa, aawiki, myaamionki* (literally 'path, time, Miami place' but described in camp materials as "Finding our paths through a Miami place, at a Miami pace, and in a Miami way"); another is *ašiihkiwi neehi kiišikwi* (earth and sky), a theme that explicitly situates land in a broader relational lens. Camp participants maintain responsibilities toward caring for and thanking the land, in recognition of how land and nonhuman relatives sustain those of us who live as humans. Participants also recognize tribal leaders, whose responsibilities include coordinating economic development for our nation to support community needs, such as funding the camps. These tribal leaders, who have been hugely supportive of cultural programs, actively engage with tribal youth, recognizing that today's youth will be future tribal leaders.

MYAAMIAATAWIAANKI KATI: RELATIONAL NARRATIVES FOR REGENERATIVE FUTURES

Colonialism disrupts relationships—between Indigenous communities and our lands, languages, cultures, and human and nonhuman relatives. Language reclamation reverses such ruptures and promotes Indigenous futures by restoring the relationships that underlie languaging. Indeed, one of the points my Miami grandfather always emphasized is the importance of strong relationships, those characterized by what a number of Indigenous scholars refer to as "*R*'s" such as *respect*, *reciprocity*, and *responsibility*. This idea is also captured in our language: *aweentii-* is a myaamia verb stem that means 'to be related to each other.' That the corresponding noun, *aweentioni*, means 'peace' captures what strong relationships can achieve. Reclamation includes not only fostering relationships between the Miami people and our language, but also our assertion of agency to define the value of those relationships. Key among our objectives is to build a positive future in defiance of the dominant discourse, whose logics almost prevented Miamis from existing as a contemporary people at all, let alone from imagining a future that centers our own culture. Clearly, we have overcome several obstacles already, but as settler colonialism is a process, so too is reclamation.

I began this chapter with an account of Miami history—both older, from the nineteenth century, and recent since myaamiaki eemamwiciki as a named narrative begins in the 1990s and continues today. I will close this chapter by sharing my brief prediction of the future. Though our language does not grammatically support "language extinction" as a concept since myaamiaataweenki is a verb and verbs cannot become extinct, it certainly provides the grammar to talk about a future. The grammatical particle *kati* marks that something will occur, and also to an extent can imply a desire for this to occur.

myaamiaatawiaanki kati. ("We will [and want to] speak myaamia.")

NOTES

1. Beyond using *myaamia* for the language, for clarity I will use *Miami* to describe the people even though the terms are interchangeable. I follow a common convention in my tribe of writing myaamia in lowercase.

2. Wesley Y. Leonard, "When Is an 'Extinct Language' Not Extinct? Miami, a Formerly Sleeping Language," in *Sustaining Linguistic Diversity: Endangered and Minority Languages and Language Varieties*, ed. Kendall A. King, Natalie Schilling-Estes, Lyn Fogle, Jia Jackie Lou, and Barbara Soukup (Washington, DC: Georgetown University Press, 2008), 23–33.

3. For an overview of community language program development, see Daryl Baldwin and David J. Costa, "Myaamiaataweenki: Revitalization of a Sleeping Language," in *The Oxford Handbook of Endangered Languages*, ed. Kenneth L. Rehg and Lyle Campbell (Oxford: Oxford University Press, 2018), 553–70.

4. Daryl Baldwin, Margaret Noodin, and Bernard C. Perley, "Surviving the Sixth Extinction: American Indian Strategies for Life in the New World," in *After Extinction*, ed. Richard Grusin (Minneapolis: University of Minnesota Press, 2018), 210.

5. Wesley Y. Leonard, "Challenging 'Extinction' through Modern Miami Language Practices," *American Indian Culture and Research Journal* 35, no. 2 (2011), 136.

6. The *Wikipedia* entry includes the *e*-word, though it ironically also mentions myaamia's "revival." See https://en.wikipedia.org/wiki/Miami-Illinois_language [accessed May 15, 2021].

7. Leonard, "When Is an 'Extinct Language' Not Extinct?"

8. I have developed the *language reclamation* framework in a number of publications including Leonard, "Challenging 'Extinction'"; Wesley Y. Leonard, "Framing Language Reclamation Programmes for Everybody's Empowerment," *Gender and Language* 6, no. 2 (2012), 339–67; Wesley Y. Leonard, "Producing Language Reclamation by Decolonising 'Language,'" *Language Documentation and Description* 14 (2017), 15–36.

9. *Revitalization* is also sometimes used to refer to what I am describing as *reclamation.* As a tenet of reclamation is that language communities have the intellectual sovereignty to name and frame their language work, my goal is not to impose a label but rather to privilege Indigenous-centered approaches, however they are named. For clarity in this chapter, however, I differentiate *reclamation* from *revitalization* as described here.

10. Kari A. B. Chew, "Weaving Words: Conceptualizing Language Reclamation through Culturally-Significant Metaphor," *Canadian Journal of Native Education* 41, no. 1 (2019), 169–70.

11. Leonard, "Challenging 'Extinction,'" 141.

12. Leanne Betasamosake Simpson, "Indigenous Resurgence and Co-Resistance," *Critical Ethnic Studies* 2, no. 2 (2016), 26.

13. I adopt for this chapter the convention of capitalizing disciplinary names (e.g., Linguistics), but using lower case to name the type of work (e.g., linguistics) that does or could occur within them. Doing so captures the critical lens of reclamation,

in this case by drawing attention to how disciplines tend to be described as if their scope is self-evident and broad (as with "Linguistics is the scientific study of language") even though they are actually characterized by particular sets of research conventions, drawing from particular (usually non-Indigenous) epistemologies and personnel. This differentiation facilitates thinking about how disciplines can change.

14. Michael Krauss, "The World's Languages in Crisis," *Language* 68, no. 1 (1992), 4.

15. Several scholars offer critiques of the extension of the biological metaphor onto languages. For example, Luisa Maffi, "Linguistic, Cultural, and Biological Diversity," *Annual Review of Anthropology* 29 (2005), 599–617; Alexandre Duchêne and Monica Heller, eds., *Discourses of Endangerment: Ideology and Interest in the Defence of Languages* (New York: Continuum, 2007); Bernard C. Perley, "Zombie Linguistics: Experts, Endangered Languages and the Curse of Undead Voices," *Anthropological Forum* 22, no. 2 (2012), 133–49; Baldwin et al., "Surviving the Sixth Extinction"; Gerald Roche, "Abandoning Endangered Languages: Ethical Loneliness, Language Oppression, and Social Justice," *American Anthropologist* 122, no. 1 (2020), 164–9. For my perspective, see Leonard, "When Is an 'Extinct Language' Not Extinct?"

16. I share an example that I found particularly violent in Wesley Y. Leonard, "Eradicating the *E*-word: Musings on myaamia Language Reclamation," *World Literature Today*, November 11, 2019.

17. Patrick Wolfe, "Settler Colonialism and the Elimination of the Native," *Journal of Genocide Research* 8, no. 4 (2006), 387–409.

18. Eve Tuck and Rubén A. Gaztambide-Fernández, "Curriculum, Replacement, and Settler Futurity," *Journal of Curriculum Theorizing* 29, no. 1 (2013), 73.

19. Joseph Errington, *Linguistics in a Colonial World: A Story of Language, Meaning, and Power* (Malden, MA: Blackwell Publishing, 2008); see also Wesley Y. Leonard, "Reflections on (De)Colonialism in Language Documentation," in *Reflections on Language Documentation 20 Years After Himmelmann 1998*, ed. Bradley McDonnell, Andrea L. Berez-Kroeker, and Gary Holton (Honolulu: University of Hawai'i Press, 2018), 55–65; and Richard A. Grounds, "The Legacy of Hunter-Gatherers at the American Philosophical Society: Frank G. Speck, James M. Crawford, and Revitalizing the Yuchi Language," in *Indigenous Languages and the Promise of Archives*, ed. Adrianna Link, Abigail Shelton, and Patrick Spero (Lincoln: University of Nebraska Press, 2021), 63–98.

20. Sylvia Wynter, "Unsettling the Coloniality of Being/Power/Truth/Freedom: Towards the Human, after Man, Its Overrepresentation—An Argument," *CR: The New Centennial Review* 3, no. 3 (2003), 257–337; see also Figueroa et al. (this volume) for a related discussion.

21. For example, Barbra A. Meek, "Failing American Indian Languages," *American Indian Culture and Research Journal* 35, no. 2 (2011), 43–60; Jenny L. Davis, "Resisting Rhetorics of Language Endangerment: Reclamation through Indigenous Language Survivance," *Language Documentation and Description* 14 (2017), 37–58; Gerald Roche, "Abandoning Endangered Languages"; Mark Turin,

"Indigenous Language Resurgence and the Living Earth Community," in *Living Earth Community: Multiple Ways of Being and Knowing*, ed. Sam Mickey, Mary Evelyn Tucker, and John Grim (Cambridge, UK: Open Book Publishers, 2020), 171–84.

22. Nancy C. Dorian, "Western Language Ideologies and Small-Language Prospects," in *Endangered Languages: Current Issues and Future Prospects*, ed. Lenore A. Grenoble and Lindsay J. Whaley (Cambridge: Cambridge University Press, 1998), 3–21.

23. Davis, "Resisting Rhetorics," 39.

24. See https://en.iyil2019.org/ [accessed September 8, 2020]. I offer a critique of the initiative's framing in Wesley Y. Leonard, "Indigenous Languages through a Language Reclamation Lens," *Anthropology News* 60, no. 5 (2019), 3–6.

25. My primary objective in drawing on this oft-repeated example is to call attention to its problematic death metaphor, but I further note that even by the colonial logics of the Endangered Languages Movement wherein a lack of first-language speakers equals "death," this statement is wrong. Research for the *Catalogue of Endangered Languages* shows the pattern of languages entering dormancy to be about one every twelve weeks. See Lyle Campbell and Eve Okura, "New Knowledge Produced by the *Catalogue of Endangered Languages*," in *Cataloguing the World's Endangered Languages*, ed. Lyle Campbell and Anna Belew (New York: Routledge, 2018), 79.

26. Jane H. Hill, "'Expert Rhetorics' in Advocacy for Endangered Languages: Who Is Listening, and What Do They Hear?," *Journal of Linguistic Anthropology* 12, no. 2 (2002), 119–33.

27. Using "speaker" rather than a broader term such as "language user" is further exclusive because it does not account for sign language users. For the current discussion, I adopt this problematic frame intentionally with recognition that the Endangered Languages Movement generally erases Indigenous sign languages.

28. Meek, "Failing American Indian Languages," 55, quoting Philip J. Deloria, *Indians in Unexpected Places* (Lawrence: University Press of Kansas, 2004), 231.

29. Davis, "Resisting Rhetorics," 40.

30. Lindsay J. Whaley, "Some Ways to Endanger an Endangered Language Project," *Language and Education* 25, no. 4 (2011), 344.

31. There is no intrinsic challenge to expressing nouns using myaamia grammar. Nouns are common, can include abstract ideas in addition to physical items, and can often be derived from verbs by adding the suffix *-oni* to a verb stem. But in myaamia, *language* isn't a noun.

32. Baldwin et al., "Surviving the Sixth Extinction," 215.

33. Leonard, "Producing Language Reclamation," 29.

34. For an overview of the camp programs from which I take these examples, see Wesley Y. Leonard and Scott M. Shoemaker, "'I Heart This Camp': Participant Perspectives on the Role of Miami Youth Camps," in *Papers of the 40th Annual Algonquian Conference*, ed. Karl S. Hele and J. Randolph Valentine (Albany: SUNY Press, 2012), 186–209.

BIBLIOGRAPHY

Baldwin, Daryl, and David J. Costa. "Myaamiaataweenki: Revitalization of a Sleeping Language." In *The Oxford Handbook of Endangered Languages*, edited by Kenneth L. Rehg and Lyle Campbell, 553–70. Oxford: Oxford University Press, 2018. https://doi.org/10.1093/oxfordhb/9780190610029.013.26.

Baldwin, Daryl, Margaret Noodin, and Bernard C. Perley. "Surviving the Sixth Extinction: American Indian Strategies for Life in the New World." In *After Extinction*, edited by Richard Grusin, 201–34. Minneapolis: University of Minnesota Press, 2018. https://doi.org/10.5749/j.ctt22nmbq0.12.

Campbell, Lyle, and Eve Okura. "New Knowledge Produced by the *Catalogue of Endangered Languages*." In *Cataloguing the World's Endangered Languages*, edited by Lyle Campbell and Anna Belew, 79–84. London: Routledge, 2018. https://doi.org/10.4324/9781315686028.

Chew, Kari A. B. "Weaving Words: Conceptualizing Language Reclamation through Culturally-Significant Metaphor." *Canadian Journal of Native Education* 41, no. 1 (2019): 168–85.

Davis, Jenny L. "Resisting Rhetorics of Language Endangerment: Reclamation through Indigenous Language Survivance." *Language Documentation and Description* 14 (2017): 37–58. http://www.elpublishing.org/PID/151.

Deloria, Philip J. *Indians in Unexpected Places*. Lawrence: University Press of Kansas, 2004.

Dorian, Nancy C. "Western Language Ideologies and Small-Language Prospects." In *Endangered Languages: Current Issues and Future Prospects*, edited by Lenore A. Grenoble and Lindsay J. Whaley, 3–21. Cambridge: Cambridge University Press, 1998.

Duchêne, Alexandre, and Monica Heller, eds. *Discourses of Endangerment: Ideology and Interest in the Defence of Languages*. New York: Continuum, 2007.

Errington, Joseph. *Linguistics in a Colonial World: A Story of Language, Meaning, and Power*. Malden, MA: Blackwell Publishing, 2008.

Grounds, Richard A. "The Legacy of Hunter-Gatherers at the American Philosophical Society: Frank G. Speck, James M. Crawford, and Revitalizing the Yuchi Language." In *Indigenous Languages and the Promise of Archives*, edited by Adrianna Link, Abigail Shelton, and Patrick Spero, 63–98. Lincoln: University of Nebraska Press, 2021.

Hill, Jane H. "'Expert Rhetorics' in Advocacy for Endangered Languages: Who Is Listening, and What Do They Hear?" *Journal of Linguistic Anthropology* 12, no. 2 (2002): 119–33. https://doi.org/10.1525/jlin.2002.12.2.119.

Krauss, Michael. "The World's Languages in Crisis." *Language* 68, no. 1 (1992): 4–10. https://doi.org/10.1353/lan.1992.0075.

Leonard, Wesley Y. "When Is an 'Extinct Language' Not Extinct?: Miami, a Formerly Sleeping Language." In *Sustaining Linguistic Diversity: Endangered and Minority Languages and Language Varieties*, edited by Kendall A. King,

Natalie Schilling-Estes, Lyn Fogle, Jia Jackie Lou, and Barbara Soukup, 23–33. Washington: Georgetown University Press, 2008.

Leonard, Wesley Y. "Challenging 'Extinction' Through Modern Miami Language Practices." *American Indian Culture and Research Journal* 35, no. 2 (2011): 135–60. https://doi.org/10.17953/aicr.35.2.f3r173r46m261844.

Leonard, Wesley Y. "Framing Language Reclamation Programmes for Everybody's Empowerment." *Gender and Language* 6, no. 2 (2012): 339–67. https://doi.org/10.1558/genl.v6i2.339.

Leonard, Wesley Y. "Producing Language Reclamation by Decolonising 'Language.'" *Language Documentation and Description* 14 (2017): 15–36. http://www.elpublishing.org/PID/150.

Leonard, Wesley Y. "Reflections on (De)Colonialism in Language Documentation." In *Reflections on Language Documentation 20 Years After Himmelmann 1998*, edited by Bradley McDonnell, Andrea L. Berez-Kroeker, and Gary Holton, 55–65. Honolulu: University of Hawai'i Press, 2018. http://hdl.handle.net/10125/24808.

Leonard, Wesley Y. "Indigenous Languages through a Language Reclamation Lens." *Anthropology News* 60, no. 5 (2019): 3–6. https://doi.org/10.1111/AN.1266.

Leonard, Wesley Y. "Eradicating the *E*-word: Musings on myaamia Language Reclamation." *World Literature Today*, November 11, 2019. https://www.worldliteraturetoday.org/blog/cultural-cross-sections/eradicating-e-word-musings-myaamia-language-reclamation-wesley-y.

Leonard, Wesley Y., and Scott M. Shoemaker. "'I Heart This Camp': Participant Perspectives on the Role of Miami Youth Camps." In *Papers of the 40th Annual Algonquian Conference*, edited by Karl S. Hele and J. Randolph Valentine, 186–209. Albany: SUNY Press, 2012.

Maffi, Luisa. "Linguistic, Cultural, and Biological Diversity." *Annual Review of Anthropology* 29 (2005): 599–617. https://doi.org/10.1146/annurev.anthro.34.081804.120437.

Meek, Barbra A. "Failing American Indian Languages." *American Indian Culture and Research Journal* 35, no. 2 (2011): 43–60. https://doi.org/10.17953/aicr.35.2.m272376nl73v332t.

Perley, Bernard C. "Zombie Linguistics: Experts, Endangered Languages and the Curse of Undead Voices." *Anthropological Forum* 22, no. 2 (2012): 133–49. https://doi.org/10.1080/00664677.2012.694170.

Roche, Gerald. "Abandoning Endangered Languages: Ethical Loneliness, Language Oppression, and Social Justice." *American Anthropologist* 122, no. 1 (2020): 164–69. https://doi.org/10.1111/aman.13372.

Simpson, Leanne Betasamosake. "Indigenous Resurgence and Co-Resistance." *Critical Ethnic Studies* 2, no. 2 (2016): 19–34. https://doi.org/10.5749/jcritethnstud.2.2.0019.

Tuck, Eve, and Rubén A. Gaztambide-Fernández. "Curriculum, Replacement, and Settler Futurity." *Journal of Curriculum Theorizing* 29, no. 1 (2013): 72–89.

Turin, Mark. "Indigenous Language Resurgence and the Living Earth Community." In *Living Earth Community: Multiple Ways of Being and Knowing*, edited by Sam

Mickey, Mary Evelyn Tucker, and John Grim, 171–84. Cambridge, UK: Open Book Publishers, 2020. https://doi.org/10.11647/OBP.0186.15.

Whaley, Lindsay J. "Some Ways to Endanger an Endangered Language Project." *Language and Education* 25, no. 4 (2011): 339–48. https://doi.org/10.1080/09500782.2011.577221.

Wolfe, Patrick. "Settler Colonialism and the Elimination of the Native." *Journal of Genocide Research* 8, no. 4 (2006): 387–409. https://doi.org/10.1080/14623520601056240.

Wynter, Sylvia. "Unsettling the Coloniality of Being/Power/Truth/Freedom: Towards the Human, after Man, Its Overrepresentation—An Argument." *CR: The New Centennial Review* 3, no. 3 (2003): 257–337. https://doi.org/10.1353/ncr.2004.0015.

Conclusion

Suzanne M. McCullagh, Luis I. Prádanos, Ilaria Tabusso Marcyan, and Catherine Wagner

This volume attends to the fact that the contemporary multiplication of extinction discourses—scientific reports quantifying species extinctions, speculative fiction on the possibility of human extinction, news articles lamenting that "[h]umans and their activities hijacked Earth"[1]—have not decelerated forces of social and ecological extinguishment. *Contesting Extinctions* diagnoses that predicament, bringing into view complex social, biological, and cultural relations obscured and erased by dominant extinction discourses. In its chapters, scholars across a range of disciplines have critically engaged with the ways in which our contemporary social and ecological moment has been materially constituted and discursively framed.

Contesting Extinctions is intended not only as a contribution to extinction studies but as an invitation to join us in confronting extinction discourses that perpetuate the impoverishment of life.[2] Contesting extinction means attending to the severing and reorganization—both material and symbolic—of ecosocial relations, changes that have brought on social inequality and biological annihilation. Contesting extinctions is a decolonial practice that reckons with colonialism's ongoing production of inequality and extinction, and calls for attention to and nurturance of regenerative, heterogeneous relations. We began the volume with the material base of such relations—land—and closed it with the symbolic ways in which we envision, alter, reproduce, and make sense of such relations—through language. In chapter 1, Leonardo E. Figueroa Helland, Abigail Perez Aguilera, and Felix Mantz reflect on how appropriate material relations with the biotic community can only be restored if stolen land is returned to its original caregivers, and in chapter 6, Wesley Y. Leonard points out that for ecosocial regeneration, it is crucial to reclaim not only land but also Indigenous languages. Matter and pattern emerge together, so we must decolonize and reclaim both the land needed to nourish our bodies and the languages that encode the knowledge for remembering, reestablishing, and reenacting our sacred relations with the land and with all our human and nonhuman relatives. Contesting extinction entails reclaiming the rights of communities to flourish within their bodies and territories in their own terms—to be able to name, feel, and honor their specific socioecological relations. Body-territories and imaginaries are sacred and complementary. A decolonial approach to extinctions and extinction discourses that centers Indigenous land-rights and linguistic reclamation will help to heal and renovate the environments and discourses[3] that have proliferated since toxic relations became institutionalized and globalized.

By way of conclusion, we draw attention to transversal threads running through and beyond *Contesting Extinctions*. To contest extinctions is, with Leonard, to contest the settler-colonial severing of relationships that render Indigenous languages and ways of life fragile.[4] To contest extinction is, with Figueroa, Perez, and Mantz, to counter forces that extinguish biological and cultural diversity and to affirm the wisdom and continuing vitality of Indigenous languages and relationships with land. To contest extinctions is, with Ryan Heryford, to interrogate the ways in which one kind of human subject is conflated with or made to stand in for the entire human species in extinction discourses. To contest extinctions is, with Lisa Ottum, to confront heteropatriarchy and affirm multispecies kinships and love. To contest extinctions is, with Brett Buchanan, to re-member the more-than-human into community.[5] To contest extinctions is, with Cooperation Jackson and Leanne Betasamosake Simpson (among others), to learn to practice community-nourishing ways of life not constrained by an orientation toward capital accumulation.[6] To contest extinctions is, with Kathryn Yusoff, to account for the violent ways in which matter and energy have been organized by racist, extractive, and imperial geographies.[7] To contest extinctions is, with Marjolein Oele, to grapple with Euro-Western linguistic impoverishment and to affirm the generative conditions for linguistic multiplicity. To contest extinctions is, with Alex Benson, to interrogate narratives and archives critically, notating and resisting discursive erasures. To contest extinctions is, with Simone Bignall, Kim TallBear, and Angela Willey, to affirm cultural complexity and engage in critical relationality.[8] To contest extinctions is, with Jodi Byrd and Mark Rifkin, to counter the temporal logics that narrate Indigenous life as past.[9] To contest extinctions is, with Nick Estes, to affirm Indigenous presence and futurity.[10] To contest extinctions is to trouble historical narratives and concepts that posit a universal humanity together in a common struggle against its future extinction.[11] To contest extinctions is to acknowledge that many humans and nonhumans have already experienced the endings of worlds.[12] To contest extinctions is, with Saidiya Hartman, Christina Sharpe, and Dionne Brand, to reckon with settler colonialism, chattel slavery, and heteropatriarchy as historical and present forces of extinguishment.[13] To contest extinctions is to expose necropolitical and racist logics underpinning Anthropocenic subject formation.[14] To contest extinctions is to acknowledge that the term "extinction" can be deployed for subjugation if "the frame of 'endangered languages'. . . inhibits reclamation," thereby "obscuring alternative futures that are crucial to reclamation praxis."[15] To contest extinctions is to retain hope by becoming, with the Red Nation, "citizens of a land that has yet to be brought into existence."[16] To contest extinctions is to reckon with the past and its legacies, thereby enabling different futures.

NOTES

1. Bridget Alex, "The Human Epoch: When Did the Anthropocene Begin?: Humans and Their Activities Hijacked Earth. Scientists Investigate When the Takeover Began," *Discover* (June 19, 2021), https://www.discovermagazine.com/environment/the-human-epoch-when-did-the-anthropocene-begin.

2. We invite the readers of this volume to explore, discuss, and use with students the audio recordings and the video from the talk by Wesley Y. Leonard for the *Approaching Extinction | Contesting Extinction* symposium (Miami University, Ohio, March 2–3, 2020), available at https://sc.lib.miamioh.edu/handle/2374.MIA/6705.

3. See the testimonies from the Gloria Quintanilla Cooperative in Nicaragua, in "The Land is Our Mother," https://friendsatc.org/wp-content/uploads/2020/10/ATC_Santa-Julia_English.pdf.

4. See also Heather Davis and Zoe Todd, "On the Importance of a Date, or, Decolonizing the Anthropocene," *ACME* 16, no 4 (2017): 770.

5. See Brett Buchanan, "Precarious Communities: Towards a Phenomenology of Extinction," in *Ontologies of Nature: Continental Perspectives and Environmental Reorientations*, ed. Gerard Kuperus and Marjolein Oele (Springer International Publishing, 2017), 219–233.

6. See *Jackson Rising: The Struggle for Economic Democracy and Black Self-Determination in Jackson, Mississippi*, ed. Kali Akuno and Ajamu Nangway (Ottawa: Daraja Press, 2017), and Leanne Betasamosake Simpson, *As We Have Always Done: Indigenous Freedom Through Radical Resistance* (Minneapolis: University of Minnesota Press, 2017).

7. See Kathryn Yusoff, *A Billion Black Anthropocenes or None* (Minneapolis: University of Minnesota Press, 2018).

8. See Simone Bignall, "Césaire and Senghor alongside Deleuze: Post-Imperial Multiplicity, Virtual Assemblages, and the Cosmopolitan Ethics of Négritude," in *Minor Ethics: Deleuzian Variations*, ed. Casey Ford, Suzanne M. McCullagh, and Karen L. F. Houle (Montreal: McGill-Queen's University Press, 2021), 245–270; Simone Bignall and Daryle Rigney, "Indigeneity, Posthumanism and Nomad Thought: Transforming Colonial Ecologies," in *Posthuman Ecologies: Complexity and Process after Deleuze*, ed. Rosi Braidotti and Simone Bignall (New York: Rowman & Littlefield International, 2019), 159–181; and Kim TallBear and Angela Willey, "Critical Relationality: Queer,

Indigenous, and Multispecies Belonging Beyond Settler Sex & Nature," *Imaginations* 10, no. 1 (July 25, 2019), http://dx.doi.org/10.17742/IMAGE.CR.10.1.1.

9. See Jodi Byrd, *The Transit of Empire: Indigenous Critiques of Colonialism.* (Minneapolis: University of Minnesota Press, 2011); Mark Rifkin, *Beyond Settler Time: Temporal Sovereignty and Indigenous Self-Determination* (Durham: Duke University Press, 2017).

10. See Nick Estes, *Our History is the Future*: *Standing Rock vs. the Dakota Access Pipeline* (New York: Verso, 2019).

11. See Yusoff, *A Billion Black.*

12. Kyle Powys Whyte, "Our Ancestors' Dystopia Now: Indigenous Conservation and the Anthropocene," in *The Routledge Companion to the Environmental Humanities*, ed. Ursula Heise, Jon Christensen, and Michelle Niemann (London: Routledge, 2017), 206–215; and Kim TallBear, "A Sharpening of the Already Present: Settler Apocalypse 2020," Department of Political Science, University of Alberta, YouTube, October 9, 2020. https://www.youtube.com/watch?v=eO14od9mlTA.

13. See Saidiya Hartman, *Wayward Lives, Beautiful Experiments: Intimate Histories of Social Upheaval* (New York: W. W. Norton, 2020); *Lose Your Mother: A Journey Along the Atlantic Slave Route* (New York: Farrar, Straus & Giroux, 2008); Christina Sharpe, *In the Wake* (Durham: Duke University Press, 2016), and Dionne Brand; *A Map to the Door of No Return* (Toronto: Vintage Canada, 2002).

14. See Stefania Barca, *Forces of Reproduction* (Cambridge: Cambridge University Press, 2020).

15. Wesley Y. Leonard, chapter 6, "Contesting *Extinction* through a Practice of Language Reclamation," 12.

16. The Red Nation, *The Red Deal: Indigenous Action to Save Our Earth* (Brooklyn: Common Notions, 2021), 29.

BIBLIOGRAPHY

Akuno, Kali and Ajamu Nangway, eds. *Jackson Rising: The Struggle for Economic Democracy and Black Self-Determination in Jackson, Mississippi.* Ottawa: Daraja Press, 2017.

Alex, Bridget. "The Human Epoch: When Did the Anthropocene Begin?: Humans and Their Activities Hijacked Earth. Scientists Investigate When the Takeover Began." *Discover*, June 19, 2021. https://www.discovermagazine.com/environment/the-human-epoch-when-did-the-anthropocene-begin.

Barca, Stefania. *Forces of Reproduction.* Cambridge: Cambridge University Press, 2020.

Bignall, Simone. "Césaire and Senghor alongside Deleuze: Post-Imperial Multiplicity, Virtual Assemblages, and the Cosmopolitan Ethics of Négritude." In *Minor Ethics: Deleuzian Variations*, edited by Casey Ford, Suzanne M. McCullagh, and Karen L. F. Houle, 245–270. Montreal: McGill-Queen's University Press, 2021.

——— and Daryle Rigney, "Indigeneity, Posthumanism and Nomad Thought: Transforming Colonial Ecologies." In *Posthuman Ecologies: Complexity and Process after Deleuze*, edited by Rosi Braidotti and Simone Bignall, 159–181. New York: Rowman & Littlefield International, 2019.

Brand, Dionne. *A Map to the Door of No Return.* Toronto: Vintage Canada, 2002.

Buchanan, Brett. "Precarious Communities: Towards a Phenomenology of Extinction." In *Ontologies of Nature: Continental Perspectives and Environmental Reorientations*, edited by Gerard Kuperus and Marjolein Oele, 219–233. Basel: Springer International Publishing, 2017.

Byrd, Jodi. *The Transit of Empire: Indigenous Critiques of Colonialism.* Minneapolis: University of Minnesota Press, 2011.

Davis, Heather and Zoe Todd. "On the Importance of a Date, or, Decolonizing the Anthropocene." *ACME* 16, no. 4 (2017): 761–780.

Estes, Nick. *Our History is the Future: Standing Rock vs. the Dakota Access Pipeline.* New York: Verso, 2019.

Hartman, Saidiya. *Lose Your Mother: A Journey Along the Atlantic Slave Route.* New York: Farrar, Straus & Giroux, 2008.

———. *Wayward Lives, Beautiful Experiments: Intimate Histories of Social Upheaval.* New York: W. W. Norton, 2020.

Leonard, Wesley Y. "Contesting *Extinction*: Toward a Praxis of Language Reclamation," talk for the *Approaching Extinction | Contesting Extinction* symposium (Miami University, Ohio, March 2–3, 2020), available at https://sc.lib.miamioh.edu/handle/2374.MIA/6705.

Red Nation, (The). *The Red Deal: Indigenous Action to Save Our Earth.* Brooklyn: Common Notions, 2021.

Rifkin, Mark. *Beyond Settler Time: Temporal Sovereignty and Indigenous Self-Determination.* Durham: Duke University Press, 2017.

Sharpe, Christina. *In the Wake.* Durham: Duke University Press, 2016.

Simpson, Leanne Betasamosake. *As We Have Always Done: Indigenous Freedom Through Radical Resistance.* Minneapolis: University of Minnesota Press, 2017.

TallBear, Kim. "A Sharpening of the Already Present: Settler Apocalypse 2020." Department of Political Science, University of Alberta, YouTube, October 9, 2020. https://www.youtube.com/watch?v=eO14od9mlTA.

——— and Angela Willey. "Critical Relationality: Queer, Indigenous, and Multispecies Belonging Beyond Settler Sex & Nature." *Imaginations* 10, no. 1 (July 25, 2019), http://dx.doi.org/10.17742/IMAGE.CR.10.1.1

"The Land is Our Mother." Testimonies from the Gloria Quintanilla Cooperative (October 2020). https://friendsatc.org/wp-content/uploads/2020/10/ATC_Santa-Julia_English.pdf

Whyte, Kyle Powys. "Our Ancestors' Dystopia Now: Indigenous Conservation and the Anthropocene." In *The Routledge Companion to the Environmental Humanities*, edited by Ursula Heise, Jon Christensen, and Michelle Niemann, 206–215. London: Routledge, 2017.

Yusoff, Kathryn. *A Billion Black Anthropocenes or None.* Minneapolis: University of Minnesota Press, 2018.

Index

About the Editors and Contributors

ABOUT THE EDITORS

Luis I. Prádanos (called Iñaki) is professor of Hispanic contemporary studies at Miami University. His research focuses on political ecology and environmental humanities in relation to Contemporary Iberian cultures. He is the author of *Postgrowth Imaginaries. New Ecologies and Counterhegemonic Culture in Post-2008 Spain* (2018). In the last years, He has guest-edited four special issues related to ecocriticism and environmental humanities for different academic journals: *Contemporary Iberian Ecocriticism and New Materialisms* (2017), *South Atlantic Ecocriticism* (2017, with Mark Anderson), *Humanidades ambientales: ecocrítica y descolonización cultural* (2019), and *Culture and Ecology in Contemporary Iberian Cultural Studies* (2019).

Ilaria Tabusso Marcyan is a visiting assistant professor of Italian at Miami University, Ohio, where she also worked as acting coordinator of The Institute for Food in fall 2020. Her research investigates how a comparative approach of Italian culture, literature, and film can help reframe contemporary environmental, ecological, social and ethical issues related to landscape, contemporary farming, and food cultures. In 2018, her article on "Slow Food and Terra Madre" (which includes a conversation with the founder Carlo Petrini) was published in the volume *Italy and the Environmental Humanities*. She is currently completing her monograph *Food Justice and Resistance in Italy: New Organic Intellectuals* (forthcoming with Lexington Books).

Suzanne M. McCullagh is an assistant professor of philosophy at Athabasca University. Her research focuses on ecological ethics, critical posthumanism, and political ecology. She is coeditor of the volume *Minor Ethics: Deleuzian Variations* (2021) and author of "Reparation Ecology and Sympathy with the Earth" in *Toward an Eco-social Transition: Transatlantic Environmental Humanities/Hacia una Transición Eco-social: Humanidades Medioambientales desde una perspectiva Transatlántica* (Biblioteca Benjamin Franklin, 2021), "Labour, Collectivity, and the Nurturance of Attentive Belonging" in *Simone Weil: Beyond Ideology*, and "Heterogeneous Collectivities and the Capacity to Act: Conceptualizing nonhumans in the political sphere" in *Posthuman Ecologies: Complexity and Process after Deleuze* (Rowman & Littlefield International, 2019).

Catherine Wagner is professor of English at Miami University in Oxford, Ohio, where she teaches poetry and creative writing. A poet and labor advocate, her writing and research address ecological and feminist issues and academic labor. Her fifth book of poems, *Of Course*, was published by Fence Books in 2020. Her previous book, *Nervous Device*, appeared from City Lights in 2012. Her poetry has been anthologized in the *Norton Anthology of Postmodern American Poetry, The Penguin Book of the Prose Poem, Best American Experimental Writing/BAX, Out of Everywhere: Linguistically Innovative Poetry by Women in North America and the UK* and elsewhere. Her writing on academic labor has appeared in *Poetry and Work* (Palgrave Macmillan, 2019), *World Social and Economic Review of Contemporary Policy Issues* and elsewhere. She is president of Miami's AAUP Advocacy Chapter, co-founder of the Save Ohio Higher Ed faculty-student coalition, and co-coordinator of Miami's Environmental Humanities Research Collaborative.

ABOUT THE CONTRIBUTORS

Alex Benson is an assistant professor of literature and American studies at Bard College, located in New York's Hudson Valley on the ancestral lands of the Stockbridge-Munsee Band of Mohican Indians. He writes about American literary history in connection with questions of sound, sociolinguistics, old media, and ecology. Recent and forthcoming essays can be found in journals including *Leviathan, Narrative*, *PMLA*, and *Small Axe*.

Leonardo E. Figueroa Helland is Chair and Associate Professor of the Environmental Policy and Sustainability Management graduate (MS) program at The New School university (Lenapehoking/Manahatta/New York City). He leads the Indigeneity, Decolonization and Just Sustainability Section of the Tishman Environment and Design Center. A decolonizing scholar of mestizo/

mix-blood heritage (Indigenous Mesoamerican and Euro-American), his work underlines the centrality of Indigenous resurgence and revitalization in addressing planetary crises, achieving climate justice and materializing systemic change. His writings appear, inter alia, in the Journal of World Systems Research, the journal Perspectives on Global Development and Technology, the volume on Social Movements and World-System Transformation, and the volumes on Anarchist Political Ecology, as well as the forthcoming NYU Environmental Law Journal (ELJ)—Special Volume on "Free the Land—Land Tenure and Stewardship Reimagined". His current projects include a manuscript prospectively titled Indigenous Resurgence beyond "Anthropocene" Collapse: From Planetary Crises to Decolonization.

Ryan Heryford is an associate professor of environmental literature in the Department of English at California State University, East Bay, where he teaches courses in nineteenth and twentieth century American literature, with a focus in ecocriticism and cultural narratives of environmental justice. He has published articles on environmental thought in the works of William Faulkner, Emily Dickinson, Mark Twain, and Édouard Glissant. His scholarship has been supported by the William Faulkner Society, the Emily Dickinson International Society, The Center for Mark Twain Studies, and the University of California Center for Global California Studies.

Wesley Y. Leonard is a citizen of the Miami Tribe of Oklahoma and associate professor of ethnic studies at the University of California, Riverside. His research theorizes sociopolitical factors that are intertwined with Native American language shift and reclamation. A collaborative project that he co-chairs, Natives4Linguistics, promotes Indigenous needs and intellectual tools as ways of doing linguistic science.

Felix Mantz is a doctoral researcher at Queen Mary University of London. His doctoral thesis examines how persisting colonial relations to land and ecologies in Tanzania are reproduced and resisted. This research is funded by the Economic and Social Research Council's LISS DTP. More generally, Felix draws on de/anti-colonial, anarchist, eco-feminist, and anti-capitalist theories as well as radical political ecology perspectives in his scholarship. His most recent work can be found in *Review of African Political Economy, Interdisciplinary Studies in Literature and Environment*, and *Review of International Political Economy.*

Marjolein Oele is professor of philosophy at the University of San Francisco. Her research intertwines ancient philosophy, continental philosophy, environmental philosophy, and philosophy of medicine. She is the author of *E-Co-Affectivity: Exploring* Pathos *at Life's Material Interfaces* (2020) and coeditor of *Ontologies of Nature: Continental Perspectives and Environmental Reorientations* (2017). Her articles have been published in a range of journals, including *Ancient Philosophy, Configurations, Environmental Philosophy, Epochê, Graduate Faculty Philosophy Journal,* and *Radical Philosophy Review*.

Lisa Ottum is an associate professor of English at Xavier University in Cincinnati, Ohio, where she teaches courses in literature, criticism, and composition. She has published on diverse authors and topics, ranging from William Wordsworth to Helen Macdonald's *H is for Hawk* to the Anthropocene.

Abigail Perez Aguilera is a decolonial feminist scholar from the Global South. She teaches at The New School, in Manhattan, NY. Her research is based on ecofeminism, Indigenous studies, environmental studies, and decolonial theory. She is a member of the Humanities for the Environment Latin American Observatory (HfE).

www.ingramcontent.com/pod-product-compliance
Lightning Source LLC
LaVergne TN
LVHW011815200926
848484LV00005B/1202